Assessments and Decisions

A STUDY OF INFORMATION GATHERING BY HERMIT CRABS

Studies in Behavioural Adaptation Series

Series Editor: John Lazarus, Department of Psychology, University of Newcastle-upon-Tyne, UK

This series examines the adaptive nature of behaviour in order to show how behaviour can be understood as a series of answers to problems set by an animal's environment. Great advances have been made in the theoretical armoury for tackling questions of behavioural evolution and adaptation, and there has been a parallel expansion in empirical studies, particularly in the field. Building on these studies, the series aims to cover a wide range of behavioural topics.

Titles published

Gulls and Plovers
The ecology and behaviour of mixed-species feeding groups
C.J. Barnard and D.B.A. Thompson

Modelling in Behavioural Ecology
An introductory text
Dennis Lendrem

Social Behaviour in Fluctuating Populations
Andrew Cockburn

Primate Social Systems
Robin I.M. Dunbar

— *Assessments* —
and Decisions

A STUDY OF INFORMATION GATHERING
BY HERMIT CRABS

R.W. ELWOOD
and
S.J. NEIL
Division of Environmental and Evolutionary Biology
The Queen's University of Belfast

CHAPMAN & HALL
London · New York · Tokyo · Melbourne · Madras

Published by Chapman & Hall, 2–6 Boundary Row, London SE1 8HN

Chapman & Hall, 2–6 Boundary Row, London SE1 8HN, UK

Chapman & Hall, 29 West 35th Street, New York NY10001, USA

Chapman & Hall Japan, Thomson Publishing Japan, Hirakawacho Nemoto Building, 7F, 1–7–11 Hirakawa-chao, Chiyoda-ku, Tokyo 192, Japan

Chapman & Hall Australia, Thomas Nelson Australia, 102 Dodds Street, South Melbourne, Victoria 3205, Australia

Chapman & Hall India, R. Seshadri, 32 Second Main Road, CIT East, Madras 600 035, India

First edition 1992

Typeset in 10/12 pt Paladium by Colset Pte. Ltd.
Printed in Great Britain by St. Edmundsbury Press, Bury St. Edmunds, Suffolk

ISBN 0 412 39110 4

A catalogue record for this book is available from the British Library

Library of Congress Cataloging-in-Publication data
Elwood, R.W. (Robert W.)
Assessments and decisions : a study of information gathering by hermit crabs / R.W. Elwood and S.J. Neil.
p. cm.
Includes bibliographical references and index.
ISBN 0-412-39110-4
1. Pagurus bernhardus — Behavior. 2. Pagurus bernhardus — Northern Ireland — Behavior. 3. Decision-making in animals. 4. Motivation in animals. I. Neil, S.J. (Susan J.), 1952- . II. Title.
QL444.M33E49 1991
595.3′842 — dc20 91-22427
CIP

Contents

Series Editor's Foreword

In the early years of this century a Scottish doctor speculated on the evolutionary origin of human tears. It seemed to him that with the increase in brain size and cognitive powers of our early ancestors many events in the struggle for existence would be just too distressing to witness. How comforting then for a mother, distraught by the sight of her child being devoured by a lion, to cloud her vision with a flood of tears! Just so, though if the good doctor had pondered further, the following picture might have occurred to him, comfortable in his speculative armchair, and given him some pause for thought.

These stories do not, of course, get us very far in explaining the evolution of tears – or anything else – but they do remind us how far the study of behavioural adaptation has come this century. This is, in fact, an exciting time for students of behaviour. The last twenty years has seen a great advance in the theoretical armoury for tackling problems of behavioural evolution and adaptation, and a parallel expansion in empirical studies, particularly in the field. The concepts of inclusive fitness and evolutionary stability, for example, have helped to explain major features of social behaviour and have generated entirely new questions for the field worker to examine. Cost–benefit analysis and optimization theory have done the same for behaviour in general, and links with population biology and population genetics are becoming stronger.

The heady days which saw the birth of behavioural ecology and sociobiology are now over, the new concepts have been refined and consolidated, and field data and comparative studies have accumulated at an impressive rate. Now seems a good time to take stock, to review the state of the art and to point some directions for future work. These are the aims of the present series, which will examine questions of behavioural adaptation and evolution in depth. As for our intended readership, we hope that all those interested in such problems, from advanced undergraduate to research worker and lecturer, will find these books of value. Some contributions to the series will encompass particular areas of study, reviewing theory and data and presenting fresh ideas. Others will report the findings of new empirical studies of an extensive nature, which make a significant contribution by examining a range of interrelated questions. The richness, but also the difficulty, of functional inquiry results from the multiple effects of selection pressures and the complex causal relationships between the behavioural responses to evolutionary forces. Studies that measure a comprehensive set of behavioural attributes, and their ecological correlates, are therefore particularly valuable.

Behaviour has functions, certainly, but has proximate causes too. Ethologists and behavioural ecologists have become increasingly aware that the decision rules of proximate causation are the means by which function is realized, and so must themselves by subject to selection. The study of the motivation and function of behavioural decisions are consequently no longer mutually divorced but instead are seeking common ground. The most ubiquitous decisions an animal makes concern resources and how to get them. To make an adaptive decision, however, the individual needs to know the value of the resource and how much it will cost to procure it. In an uncertain world this puts a premium on gathering information and assessing its relevance, so that both motivational and functional inquiry must consider the cost and value of information gathering for a full understanding of behavioural decision making.

Addressing these issues here Robert Elwood and Susan Neil call on their own, and their coworkers', extensive studies of hermit crabs to uncover the subtlety of the information-gathering process. The shell that a hermit crab inhabits is its most precious resource but the suitability of shells varies enormously and can only be assessed by visual and tactile means. Hermit crabs have therefore evolved special behaviour patterns and associated decision processes to assess the quality of a shell that might be adopted. The unusual and intriguing aspect of shells as resources is that in assessing their suitability the crab must consider the advisability of a move from its present abode to a new one: a comparison process is crucial if a move is to be adaptive. When the potential new shell is already occupied by another crab the crab must assess, in addition, the likelihood and costs of wresting it from its occupier in a competitive encounter. Our current understanding of these assessment processes is thoroughly reviewed in these pages, revealing both the scope and the limitations of the hermit crab's achievement, and presenting some ingenious and elegant experiments in the process. The accounts of related work on other species with other problems set the hermit crab experiments and models in their wider context and demonstrate the importance and generality of the problems of assessment and decision making.

JOHN LAZARUS

Preface

This book is the result of many years of collaboration which began after a chance meeting. Bob had been studying hermit crabs for several years in Belfast when he took a sabbatical year (1981–2) to work in the University of Minnesota. Sue was in the adjacent office at Minnesota, just completing a PhD on field studies of aggression in cichlid fish. Sue had been awarded a NATO fellowship to work with Geoff Parker in Liverpool on aggression in hermit crabs. We thus discovered our mutual interest in these animals. Sue took up her fellowship in Liverpool but crossed the Irish Sea to visit the animal behaviour laboratory in Belfast. She subsequently moved to Belfast, initially just for a few months, but remained for several years to continue our joint work on decision-making in hermit crabs. Our collaboration brought together Sue's background in functional analyses of behaviour with Bob's interest in causation.

Our work has been helped by many people to whom we are greatly indebted. Thanks are due to three postgraduate students supervized by Bob; Anne Stewart (nee McClean), Barbara Dowds and Nicholas Jackson. Thanks are also due to other students who participated in projects involving hermit crab assessments and decisions; Linda Webb, Anne French, Christopher Glass, Avril Watson, Valerie Stamford, Stephen Preston, Nicola Marks, Pauline Adams and Patricia Duffy. Thanks go to Hazel Kennedy, who worked as a research assistant. We have discussed our work with Geoff Parker, Robin McCleery, Ernie Reese, Brian Hazlett, Peter Abrams, George Barlow, Donald Broom, David McFarland, John Krebs, Viv Grotto, Michio Imafuka, Frank McKinney and Jaimie Dick who, knowingly or not, all provided help for which we are grateful. Thanks are also due to Paul Miller for rearranging sections of the text to suit the dictates of the English language. Finally, special thanks are due to John Lazarus who meticulously read the entire text and whose thoughtful comments resulted in considerable improvements to the clarity of our arguments.

BOB ELWOOD and SUE NEIL

CHAPTER 1

Assessments and decisions

'In the vast literature dealing with the selection of nest sites, nest building materials, spawning sites, prey, mates, etc., there has been relatively little speculation as to the nature of the mechanisms underlying selection or choice behaviour' (Reese 1963). Although this statement was made over a quarter of a century ago it still has an element of truth. Much has been discovered concerning the resources that animals prefer and the stimuli to which they respond, but relatively little is known about the methods of assessment employed by animals or about the underlying motivational basis of their decisions.

The terms 'assessment' and 'decisions', as used above, are common in recent ethological literature. The former term became popular after game theory was used in the analysis of animal contests (Maynard Smith 1974). An 'assessment strategy' involves an inequality or inequalities being detected by the participants in a contest and determining the eventual winner (Parker 1974). But animals can also be said to make assessments in contexts other than aggressive encounters. For example, female robins (*Erithacus rubecula*) may assess the size of a male's territory and choose mates on that basis (Harper 1985), female mallard ducks (*Anas platyrhynchos*) may assess male quality (Holmberg *et al.* 1989), male *Gammarus* assess female quality both prior to and during precopulatory amplexus (Hunte *et al.* 1985; Dick and Elwood 1989), male tiger blue butterflies (*Tarucus theophrastus*) assess the number of potential competitors (Courtney and Parker 1985) and assessments of costs and benefits are made by animals when foraging (Krebs and McCleery 1984; Godin and Rangeley 1989). The term 'assessment', therefore, clearly has its ethological application in the functional analysis of behaviour. Occasionally the term 'evaluate' is used in place of 'assess' (e.g. Wetterer 1989).

In contrast, the word 'decision' comes primarily from motivational analyses (McFarland 1971; Dawkins and Dawkins 1973). It has been used to describe a process by which behavioural change is non-random; that is, given a number of potential activities, some are more likely to occur than others in any given set of circumstances. We should distinguish the process by which the activity is selected from the occurrence of the activity itself, referring to the former as 'decision-making' and the latter as the 'result of the decision'. The decision-making mechanisms are assumed to have been subject to natural selection and thus the resultant decisions are expected to be economically sound in that they will maximize the fitness of the individual (Sibly and McFarland 1976; McCleery 1978).

The concepts of 'assessments' and 'decisions' clearly have features in common. Assessments form the basis of some decisions, but as we shall see, not all decisions are based on assessments.

DEFINING ASSESSMENT

If one observes two animals of the same species responding to similar stimuli their responses may be different. This difference may be of a quantitative nature, with the same activities shown but with different durations or vigour, or it may be qualitative, different activities being shown. It is possible that two animals respond differently to similar stimuli because they have assessed the situation differently. For example, in many species fighting is influenced by the potential costs and benefits involved (e.g. Parker 1974; Thornhill 1984). Thus one animal may attack because it is larger than the opponent and another animal may retreat because it is smaller. Alternatively, an animal may be more likely to attack an opponent of any size if the potential gain from the contest is great. Much variability in behaviour therefore is attributable to assessments but much is caused by other factors and, in order to discuss assessments, we should define what is meant by that term. To facilitate this task behavioural variation is classified below according to the developmental route by which it arose, thus distinguishing variation that arose due to assessments from that which did not. The classification, however, is not intended to be a comprehensive categorization of ontogenetic origins of behaviour.

The main dichotomy in the classification concerns whether or not responses differ because each individual has a fixed or varied mode of response. In practice, all animals show variability in response but often the differences between individuals are more dramatic than those within individuals. It is these inter-individual differences that are considered first.

Inter-individual variation

In this section permanent or relatively permanent behavioural differences between individuals are discussed. These differences may be caused by either genotypic or environmental effects. It is not suggested that either of these alone causes a particular activity, but that they are major factors influencing behavioural differences.

(a) Genotypic effects

That genetic differences influence behavioural differences is most easily demonstrated by the study of inbred or mutant strains and recent reviews have illustrated numerous and varied examples (e.g. Partridge 1983). Genetic differences may also be evident when examining individuals that have been drawn from different natural environments and are thus subject to different selection pressures. For example, sticklebacks (*Gasterosteus aculeatus*) from different habitats differ in the degree of fright response shown to models of avian predators (Giles and Huntingford 1984) and this appears to be correlated with the level of avian predation in those different habitats. Since offspring of the two populations reared in the laboratory also show this difference it must be at least partly due to a genetic difference between the two populations (Giles 1984). It seems that there is an optimal fright response in different habitats and that separate populations each approach their own optimum. Within one habitat, therefore, selection may reduce variability and there is no need to assume that sticklebacks assess the level of avian predation.

In contrast, there are examples of behavioural variability that are maintained within a single population by frequency-dependent selection. For example, Gross (1985) has reported that male Pacific coho salmon (*Oncorhynchus kisutch*) exhibit two distinct morphological and behavioural phenotypes and suggests that these have separate genetic determination. Some males become sexually mature at 2 years of age (jacks) whereas for others maturity is delayed until 3 years (hooknose males). The large hooknose males compete for access to females during spawning by fighting and the largest males perform best in this form of competition. The small jacks, on the other hand, tend to obtain access to females by sneaking up to spawning sites using rocks as hiding places. The smallest males do best at this strategy. Medium size males do badly at both strategies, thus providing the basis for disruptive selection. Gross calculated the lifetime fitness of jack and hooknose males for the observed frequencies of occurrence and found these to be virtually identical, which suggests frequency-dependent selection. There is no evidence to suggest environmental determination of the two phenotypes or that they reflect a 'best of a bad job' strategy (Gross 1984). Breeding experiments are required, however, to establish firmly that genotypic differences control the behavioural differences. There is no evidence at present that assessment influences these differences.

(b) Environmental effects

Salmon are not the only animals to show two distinct male phenotypes. Curious examples occur in parasitic copepods and in these cases the phenotypes have been demonstrated to be environmentally determined. One such species (*Pseudomyicola spinosus*) lives on the gill filaments of the Japanese blue mussel (*Mytilus edulis galloprovincialis*). One male form is a small active swimmer, whereas the other form is larger and capable only of crawling (Do *et al.* 1984). If single larval males are introduced separately onto sections of gill filament, 96% of them develop into active swimmers. In contrast, if several larvae (male and female) are introduced onto one section of gill filament, then 97% of the males develop into the larger crawling morphs. Each male thus attempts to maximize the chances of locating females. Other copepods also show male dimorphism (Hipeau-Jacquotte 1984).

Naturally, many examples are known of environmental effects having permanent or relatively long-lasting influences on behaviour. Nutrition can be an important factor, for example in influencing the behaviour of queens and workers in the social Hymenoptera (Wilson 1971). Nutritional effects can also have long-lasting effects in mammals (Smart 1983). The intrauterine environment of the embryos may also have dramatic effects on subsequent behaviour of mammals. In rodents, for example, the sex of the neighbouring embryos influences the hormonal environment and this appears to influence later behaviour (vom Saal 1984). Exposure to a variety of external stimuli can have long-lasting effects in many types of animals, particularly with respect to social preferences (Chalmers 1983; Slater 1983).

It is clear that where genotypic differences are responsible for behavioural variation no assessment takes place. However, when the environment influences a permanent change some 'assessment', broadly defined, must be going on. Thus in the case of parasitic copepods the males presumably detect (i.e. assess) the presence or absence of females within a host and 'decide' upon the appropriate form. The term 'life-history decision' may be used to mean the commitment of a period of time to a particular course of action (Brockman *et al.* 1979). Of course, conscious decisions are not implied, but in some cases it is apparent that information gained from the environment at one time influences the subsequent life of the animal. However, in this text the term 'assessment' will be restricted to cases in which the subsequent decision does not result in a permanent morphological or behavioural phenotype. It is thus restricted to cases of intra-individual variation.

Intra-individual effects

It is when the behaviour of individuals varies with time that we typically find assessment strategies. Assessment is not, however, involved in all cases of intra-individual variation.

(a) Fixed, mixed evolutionarily stable strategy (ESS)

Individuals may employ a genetically determined fixed, mixed strategy wherein the behaviour varies from time to time but the probabilities of occurrence for each component of the strategy are not influenced by environmental stimuli. This is described as an evolutionarily stable strategy (ESS) in which the utility of each of the component parts is the same (Maynard Smith and Price 1973). A theoretical example is the hawk/dove game (Maynard Smith 1974). In this game each contestant plays one strategy with probability p and the alternative with probability $(1-p)$. No assessment of the environment or of the opponent occurs to influence these probabilities. A practical example comes from a study of the digger wasp *Sphex ichneumoneus* (Brockman and Dawkins 1979; Brockman *et al.* 1979). Females of this species show a mixed strategy for obtaining burrows in which to lay their eggs, either digging a burrow of their own or appropriating one already constructed. Some of these ready-made burrows have been abandoned, but others are being used by other females. If a wasp enters one of the latter type she may provision it with food only to have the other female lay her eggs and close the burrow. It may therefore be advantageous for digger wasps to enter abandoned burrows but to avoid those in use by other females. Females do not do this, however, presumably because they cannot distinguish between a burrow that has been abandoned and one that has not. Thus no assessment is possible.

(b) Maturational change

An animal's behaviour may change with different stages of development – for example, in relation to age or size. These changes are fixed in the sense that they are not dependent upon information about the environment. For example, Butler (1949) describes changes in the behaviour of worker honey bees (*Apis mellifera*) which, after emerging as adults, continue to show physiological and behavioural changes. In the first three days after pupal emergence the workers solicit food from other workers and restrict their 'work' to the cleaning of brood cells. In the following few days workers start to take food from stores within the hive and also commence feeding old larvae. Subsequently they feed young larvae, and then progress to guard duties. Only after three weeks have elapsed do workers begin foraging outside of the hive. These changes in behaviour appear to be determined by age-dependent physiological changes.

Other examples of changes occurring in a fixed manner are to be found among sequential hermaphrodites. The age or size at which sex change and thus behavioural change occurs is set by natural selection in such a way that the overall success due to the male and female functions is maximized (Charnov 1982), taking into account the normal mortality rates of the two sexes. An example of this is seen in the shrimp *Pandalus borealis*, a protandrous sequential hermaphrodite (changes from male to female). When a population of this species was fished commercially it was the older, larger females that suffered the higher mortality. Thus there were fewer females per male and the utility of remaining male decreased. In subsequent years the size at which sex change occurred was reduced. Charnov (1981) presumed this to be due to selection changing the ESS size for optimal sex change (see also Warner 1975) and there was no suggestion that assessment was involved. Not all sequential hermaphrodites, however, conform to a fixed maturational sequence; some respond to environmental events, particularly with respect to dominance relationships (Fricke and Fricke 1977). These latter changes cannot be classed as following a fixed maturational 'strategy' but instead are environmentally induced permanent changes (see Environmental effects, p. 3).

(c) Assessment

The key feature of assessment is that some aspect of the environment is perceived and the information used as a factor in decision-making. Neither the fixed, mixed ESS nor many maturational changes show this feature of using information from the environment in decision-making. It is suggested above that those cases in which an environmental change results in a maturational change should not be classed as involving an assessment. Many examples of assessments are given elsewhere in this text, but here we mention one case which should help to clarify the distinction between assessment and maturation.

Males of different sizes show different mating strategies in a number of species; for example, some anuran males either call to attract females (callers) or remain silent and attempt to intercept females that approach callers (satellite or sneaker males). This dichotomy has been referred to as a strategy conditional on phenotype (Davies 1982), a classification that incorporates Environmental effects (p. 3) and Maturational change (p. 5). Indeed if they were following the strategy 'if small sneak; if large call' they would fulfil the conditions of one of those two classifications. Which one was fulfilled would depend on whether the animal changed size between reproductive attempts, i.e. the response was one stage in a maturation process (p. 5), or whether its size remained constant, i.e. it was a fixed life-long response (p. 3). There is some evidence to suggest that *Rana catesbeiana* conforms to a maturational strategy; that is, the probability of calling is size-dependent (Howard 1978). The green tree frog (*Hyla cinerea*) also shows a significant correlation between body size and the probability of calling (Perrill *et al.* 1982). In this latter case, however, males are sensitive to what others in the population are doing; that is, they assess whether others are calling and this influences the probability that they will call (Perrill *et al.* 1982). Thus some of the behavioural variation may be attributed to maturational change and some to assessment processes (see also Arak 1988).

The term 'assessment' needs to be defined precisely as it has been used in a number of different ways in the past. Parker and Stuart (1976) define 'resource assessment' as 'a process in which resource cues correlating with expected gain in fitness are monitored and which results in adaptive motivational changes'. They go on to suggest that 'assessment' means 'the cue-monitoring process and the decision'. There has, however, been a more recent tendency to use assessment as meaning the cue-monitoring process alone and to distinguish it from the decision (e.g. Dowds and Elwood 1983; Harper 1985; Simbolotti *et al.* 1987). There is therefore a tendency to talk of assessments being used as the basis for decisions and this is the sense that is employed in the present text. 'Assessment' may be used to mean 'gathering information' (Elwood and Stewart 1985; Elwood and Neil 1986a), though this latter term implies an active process, perhaps entailing the animal in some behavioural effort or work in order to obtain the relevant information. This is the case in our hermit crab studies, and in many other examples, in which time and effort are expended by the animal in the assessment process. In many cases of apparent assessment, however, it is difficult to state whether or not time and effort has been expended to gather information and the term 'cue-monitoring' is preferred as it does not have this connotation. The use of the term 'assessment' thus implies clearly that information is gathered which results in motivational change. It is because of this emphasis on motivational change that 'life-history decisions' are not regarded here as being influenced by assessments.

(d) Physiological change

An individual may vary in its response to a particular environmental situation because of a change in its internal state. The response to food or water may depend largely on the nutritional or hydrational state (Toates 1980) and the response by a male to a potential mate may be influenced by the amount of sperm he has available (Halliday 1976). Virtually all behavioural systems are influenced by the animal's internal state such that the same animal will respond to the same set of external stimuli differently at different times.

Several physiological variables may influence the tendency to perform any particular activity. For example the hydrational state of certain hypothalamic cells, the blood pressure on the venous return to the heart and the renal arteries, and the level of angiotensin in the plasma all appear to influence the tendency to drink (reviewed by Toates 1980). We may refer to each variable that is monitored by the animal as a separate causal factor

(McFarland and Sibly 1975). All internal factors that are relevant to potential behavioural acts make up the 'physiological state' (e.g. McFarland and Sibly 1972; Sibly and McCleery 1976) and there will be an optimal state for the animal, i.e. a thermal, hydrational and nutritional state, etc. that allows for the most efficient biochemical and behavioural activity. It is important therefore that the animal confines the variability of each internal factor within tolerable limits and thus avoid states that are likely to be detrimental to its fitness. To this end, the internal state should be monitored and behaviour influenced in such a way as to minimize detrimental variation of the physiological state.

ASSESSMENT AND PHYSIOLOGICAL STATE

We have discussed terms of assessment and physiological state as though they were separate concepts, and this is how they have been treated in many texts. Assessment is usually discussed under the heading of function whereas physiological state, along with the concept of decisions, under the heading of causation. Recent trends, however, have moved away from this position with the realization that the way internal factors influence decisions has been subject to selection and that decisions should therefore be 'optimal' in the same manner as assessments (McCleery 1978; McFarland and Houston 1981). Furthermore, the process of making assessments must involve a change in the motivation of the animal and this change may be described in terms of causal factors and causal factor space in the same manner as are physiological states (McFarland and Houston 1981). If this is the case the two concepts may be difficult to distinguish in practice. Certainly shifts in responsiveness of animals may be as a result of changes in both internal and external causal factors.

A well-known example comes from a study of courtship in the guppy, *Poecilia reticulata* (Baerends *et al.* 1955). Males vary through time in their tendency to court a particular female and this has been described as a change in 'libido' or internal state. Particular colour patterns on the males make it easy to judge their internal state. Males also show variation in their tendency to court different sizes of females, with large females (capable of producing more offspring) being much more likely to be courted than small females, which indicates that female quality is assessed. The overall tendency to court is a product of the internal state and the size of the female (Fig. 1.1). It is not possible to state that courtship in any one instance is due only to either the internal state of the male or to his assessment of the female.

Courtship may be influenced by factors directly related to the male's 'libido' as we have seen above, but it may also be influenced by physiological factors not normally associated with courtship. In newts the number of spermatophores a male has influences his 'libido' and the responsiveness of

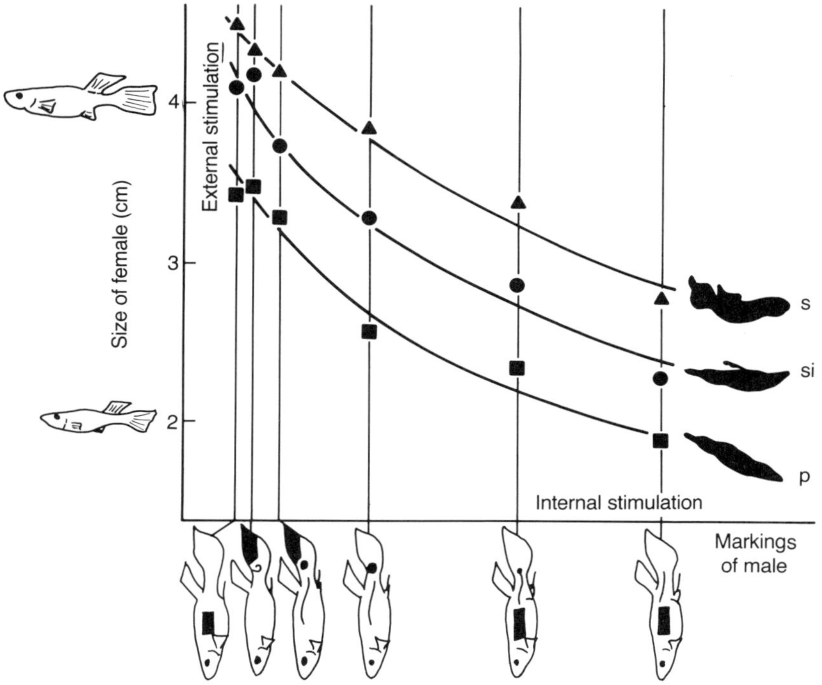

Figure 1.1 The influence of the strength of external stimulation (measured by the size of the female) and the internal state (measured by the colour pattern of the male) determines the courtship behaviour of male guppies. Each curve represents the combination of external stimulus and internal state producing posturing (p), sigmoid intention movements (si), and the fully developed sigmoid (s), respectively. (From Baerends *et al.* 1955.)

the female influences his tendency to court (Halliday 1977). Changing the oxygen content of the water, however, also has marked effects on courtship, with decreasing oxygen levels causing a more rapid courtship and increasing levels causing a longer courtship. Thus the behaviour is determined by a variety of internal and external factors that interact in complex ways. The important point here is that even if animals make the same assessment, in this case by gathering the same information about the responses of the female, the decisions may be different because of differences in the internal state.

It has also been suggested that the assessment itself may be modified by physiological state. For example, in asymmetrical contests over a disputed resource it is predicted that the costs and benefits of the potential dispute should be assessed in some way and used to determine the winner (Parker

1974). If food is the resource, the value that any one animal places on acquiring the food will depend, in part, on the nutritional state of the animal. Experimental evidence of changes in aggression over food has been found in water-striders (Wilcox and Ruckdeschel 1982), in which aggressiveness and consequent territoriality are determined to a large extent by nutritional state. The assessed value of food is therefore related to the physiological state of the animal.

DISCRETE AND CONTINUOUS ASSESSMENTS

An assessment may result in an animal deciding among two or more discrete activities or among different durations or intensities of one activity. It is possible that in the former case the animal discriminates between two discrete environmental variables and decides to respond appropriately to the situation. A male might, for example, discriminate between virgin females and mated females, courting the former but rejecting the latter. It is also possible, however, that separate courses of action may follow from an assessment of a continuous variable. Thus, in *Gammarus*, males assess how close a female is to moulting (at which time mating can take place) and decide whether to take the female into a precopulatory guarding phase or to reject her on this basis (Elwood *et al.* 1987). Thus the observed decision is discrete but the assessed variable is continuous. When the decision is continuously variable we would anticipate that the feature(s) assessed is also continuously variable (Fig. 1.2). An example of this is the influence of relative size on the decision regarding fight durations in scorpionflies (Thornhill 1984). The closer in size are the two contestants the longer the fight.

In the remainder of this section we examine the basis of discrete and continuously varying behavioural outcomes with particular reference to reproduction in parasitic Hymenoptera. Female parasitic Hymenoptera lay their eggs on living, paralysed or dead hosts and have the ability to deter-

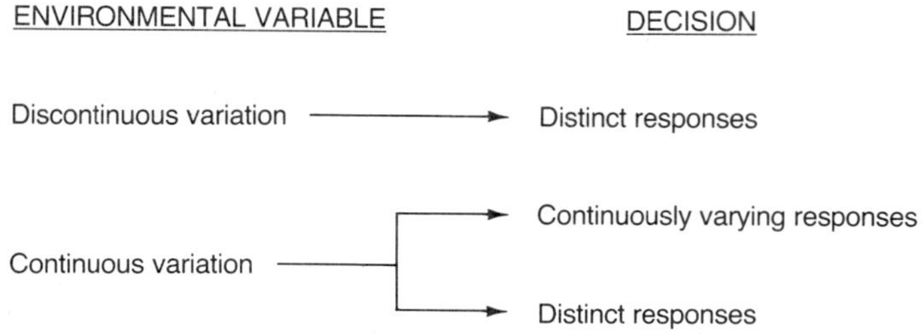

Figure 1.2 Discontinuous variation in the environment may be assessed and result in distinct responses whereas continuous variation tends to result in either distinct responses or continuously varying responses.

mine the sex of their offspring. Eggs that are unfertilized, and therefore haploid, are male whereas eggs that are fertilized, and therefore diploid, are female. The adult female can control whether or not her eggs are fertilized, and hence their sex, by the use of sphincter muscles retaining the sperm in her stores. In cases of a single parasite being produced per host the size of the emerging adult may be correlated with the size of the host. The reproductive success of each sex is usually correlated with body size but sometimes the success of females is more strongly influenced by size than is that of males (see Charnov 1982 for review). This is the situation with *Heterospilis prosopoidis* which attacks bean weevils (Charnov *et al.* 1981) and *Lariophagus distinguendus* which attacks granary weevils (Assem 1971; Charnov *et al.* 1981). An adult female would attain a higher fitness by laying female eggs on the larger hosts and male eggs on the smaller than she would by laying eggs at random. In order to avail herself of this higher fitness, however, the female must possess the abilities to assess host size and to vary the sex of her offspring accordingly.

If female *Heterospilis* are exposed to hosts of a single age (and hence size) they produce a male-biased ratio (0.85 males) when the hosts are 11 days of age but a female-biased ratio (0.4 males) when the hosts are 13, 15, 17 or 19 days. Charnov *et al.* (1981) found no difference in the sex ratio produced in the four older age classes but a major difference between ages 11 and 13 days. This suggests a simple threshold in the assessment of host size that determines the probability of producing a male.

Female *Lariophagus* on the other hand show an ability to produce a continuous range of sex ratios from all male down to only 15% males depending on host size (Fig. 1.3) (Assem 1971; Charnov *et al.* 1981). Furthermore, when two host sizes were offered the sex ratio produced on each depended on the relative sizes rather than the absolute sizes. For example, when hosts of 1.4 mm (host size measured as the width of the tunnel in the grain) were offered alone about 15% of the offspring were males. When this size was paired with a larger size more males were produced on the 1.4 size hosts (30% males). When offered with a smaller host size the sex ratio switched to about 2% males (Fig. 1.3). These data indicate not only an ability to make assessments along a continuum but also suggest that the assessment is dependent, in part, upon a memory of previously encountered hosts (Green 1982). It should be noted, however, that each separate decision is discrete (i.e. to produce a female or male egg) but that these decisions are based upon an assessment of a continuous variable. The mechanisms involved in this assessment and the decisions made concerning egg deposition in *Lariophagus* are quite complex and are considered more fully towards the end of this chapter.

The utility of producing male as opposed to female eggs also changes when male offspring are likely to compete amongst themselves for the opportunity to mate, a phenomenon known as local mate competition (Hamilton 1967).

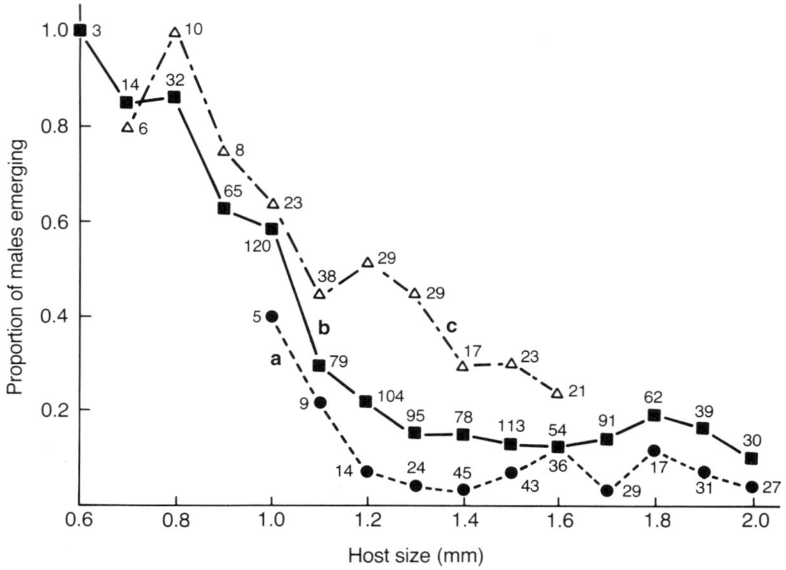

Figure 1.3 Sex ratio in *Lariophagus*. The proportion of males emerging is plotted against the size of the hosts (weevil larvae) (tunnel width) which were offered in a 'carousel experiment'. Curve b results when female wasps are presented sequentially with 20 hosts of a single size. Curves a and c result when wasps are presented with an alternating sequence of two host sizes, where the hosts are 0.4 mm different in size. Curve c is where the host of interest was the smaller of the two. For example, hosts of 1.4 mm were offered alone and gave a sex ratio of 15% males (curve b). When they were offered alternately with 1.8 mm hosts, they gave a sex ratio fo 30% males (curve c). When offered with 1.0 mm hosts, their sex ratio dropped to 2% males (curve a). The number of emerging wasps is given for each data point. (From Charnov *et al.* 1981.)

This occurs in *Nasonia vitripennis*, the females of which attack fly larvae and typically lay 10–50 eggs of which 10–15% are males. The wingless males hatch first and mate with their sisters as the sisters emerge, the few males being sufficient to mate with all the females (Werren 1980, 1983). The female thus avoids producing wasted males but this does not require assessment, just a fixed rule. However, if females encounter a larva that has already been parasitized, the female appears to assess this, lays eggs in the parasitized host (= superparasitism) but typically produces a few male eggs only. A discrete assessment and decision occurs. If hosts are clumped and are utilized by other females such that males from one brood may mate with females from another the conditions for local mate competition break down and it would benefit a female to produce a ratio nearer to 50% males. This has been shown to occur (Werren 1983), but this is not due to the superparasitism noted

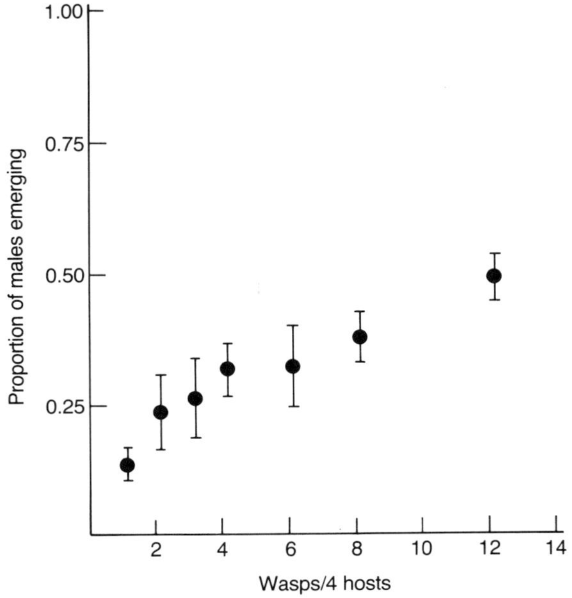

Figure 1.4 Sex ratio from patches containing 1 to 12 associated ovipositing females for the wasp *Nasonia* attacking blowfly pupae. Mean values ±2 standard errors are represented. (From Werren 1983.)

above. Rather, it appears to be due to the number of females ovipositing in a particular patch being monitored, i.e. assessed, by each female (Fig. 1.4). As the number of ovipositing females increases so does the proportion of male eggs produced suggesting both assessment and decisions along a continuum.

Studies on another parasitoid (*Trichogramma evanescens*) provide similar findings, but more importantly, a cautionary tale. Females of this species assess host size and lay more eggs in large hosts (Klomp and Teerink 1962). They can also assess whether or not the host has been parasitized and avoid superparasitism (Salt 1937). Hosts are often clumped and a single female laying eggs on a clump produces a female-biased sex ratio (15–35% males) in accord with the theory of local mate competition (Hamilton 1967). When other females are present, however, the sex ratio approaches equality, as in the studies on *Nasonia*, and attempts have been made to discover the nature of this apparent assessment (Waage and Lane 1984; Waage and Ng 1984). Could females assess the population density by encounters with other females or could they assess parasitized hosts by some individual marker and use this to determine the number of other females? The data suggest that neither of these features is assessed by *Trichogramma*. Rather, these parasitoids use a single fixed rule. Females producing a sequence of eggs tend to produce males within the first few eggs and then only at set intervals thereafter. Thus, females producing a few eggs produce approximately 31% males, whereas those producing more eggs have a smaller proportion of males: 21%. Females encountering single hosts in a sequence, each separated by a 30-minute interval, produce approximately one male per host. Those encountering ten hosts together produce one male on the first host but fewer thereafter (Fig. 1.5). Thus the shift to a higher proportion of males when more females are present is not due to an assessment procedure, rather it is by a simple fixed rule based upon interruption of egg-laying. This illustrates

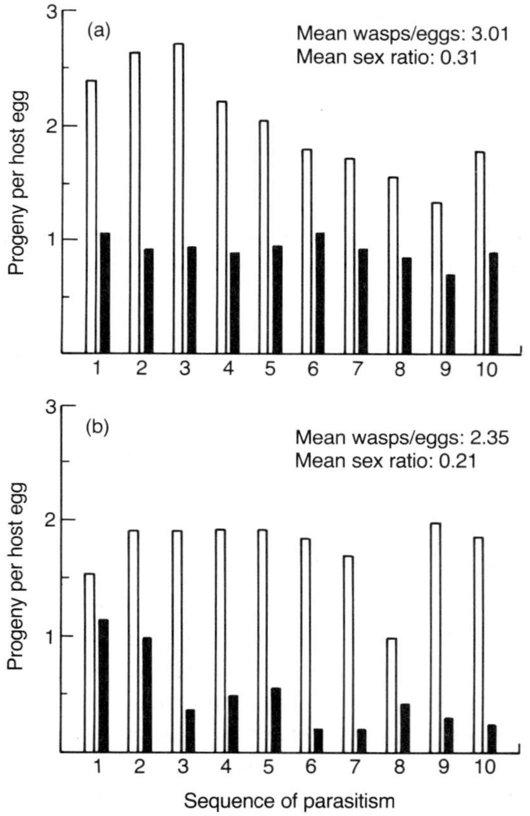

Figure 1.5 Average numbers of females (white) and males (black) produced by *Trichogramma* on ten host eggs offered in sequence. (a) Host eggs offered singly, at 30-minute intervals (x = 1). (b) Host eggs offered in a single mass (x = 10). (From Waage and Ng 1984.)

the need to be aware of more parsimonious explanations than those provided by inferring 'assessments.

The above discussion should not lead to the conclusion that *Trichogramma* is incapable of assessment because, as stated, females lay more eggs on large hosts than on small hosts. How then is this size assessment achieved? Experiments on *T. minutum* have indicated that olfactory and visual stimuli are not required (Schmidt and Smith 1987). Before laying eggs, however, the female walks over the host, first directly across the host and, on reaching the other side, making an abrupt turn and continuing apparent investigation. Only one feature of this walking was found to correlate with the number of eggs produced by the parasite and that was the time taken in the initial transit across the host. Time was shown to be the basis of the assessment by the placing of a barrier across some hosts. In these cases the initial walk terminated prematurely on reaching the barrier and fewer eggs were produced than when no barrier was present. That the important feature is time and not number of steps taken was inferred from a comparison of the responses of large and small wasps. The former took fewer steps but the same time as the latter and they did not differ in the number of eggs produced. This provides a clear example of mode of assessment of a continuous feature (size) leading to a continuous decision (number of eggs).

DECISION-MAKING

It is apparent that when an animal stops performing one activity it has to decide or has already decided which activity to perform next. This is a clearly visible 'decision point'. The animal must also decide when to switch from one activity to another and it is expected that animals somehow review the situation as regards the internal and external state and on that basis decide whether or not to continue the current activity. If the decision is to continue the current activity then this decision will not be visible to the observer. We should thus seek a model of decision-making that explains a continuous process in which most of the decisions are invisible and only those involving behavioural change are visible.

At any one time causal factors for several activities are likely to be present; how then does the animal select which activity to perform? A common sense solution is to suppose that all causal factors are 'weighted' as regards their consequences for the fitness of the animal and for the animal to select that activity for which there is the highest level of causal factors. Such a model, presented by McFarland (1969), illustrates two extremes of behavioural change, these being termed 'competition' and 'disinhibition' (Fig. 1.6). When takeover is by competition (Fig. 1.6a) the timing of that takeover is caused by the increasing level of causal factors associated with the new activity. If these start at a low level (B_x in Fig. 1.6a) then behavioural change will be

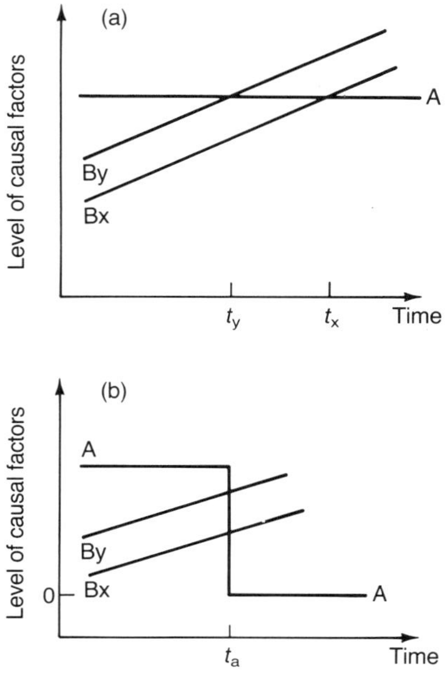

Figure 1.6 Level of causal factors as a function of time. In (a) behaviour B ousts activity A by competition. In (b) B is disinhibited by A, as explained in the text. t_y, t_x and t_a indicate the timing of transition from behaviour A to behaviour B. (From McFarland 1969.)

later than if they started at a higher level. When takeover is due to disinhibition, however, the time of takeover is influenced by the level of causal factors associated with the first activity (Fig. 1.6b). When these causal factors decline the activity that is second in priority is disinhibited. The level of causal factors associated with this second activity does not influence the time of its onset, although it might influence the duration and vigour of that activity (Rowell 1961; McFarland and L'Angellier 1966). The main predictions of this model have been tested in an elegant study of decisions by male newts on whether to court females under water or to breathe air at the surface (Halliday and Sweatman 1976). Breathing was disinhibited if the female was suddenly removed from the vicinity of the male or if the courtship went to completion; however, if neither of these occurred breathing eventually took over by competition. Competition or disinhibition were inferred from the predictability of the timing of the onset of breathing.

Disinhibition may occur in times of conflict between two opposing behavioural tendencies and result in the occurrence of displacement activities (van Iersel and Bol 1958). If the 'expected' result of an activity fails to occur or is blocked, disinhibition may occur because of the animal 'switching attention' (McFarland 1966, 1969). It may also occur when the first priority activity ends because of satiation or if the goal is reached. Finally there may be specific mechanisms of disinhibition relating to certain behavioural systems (McFarland 1969). It is anticipated that natural selection has shaped particular behaviour sequences so that they can satisfy particular needs. For example in many animals vigilance is interspersed with feeding bouts, and the former may be a disinhibited activity. Again this might be due to an attention-switch that has the effect of suddenly reducing the causal factors associated with the first activity allowing the second priority activity to show through for a short period (McFarland and Lloyd 1973; Cohen and McFarland 1979). This 'time sharing' hypothesis has recently been criticized, however (Crawford and Colgan 1989).

A different approach to decision-making is used in the state space model (McFarland and Sibly 1975). This approach assumes that all activities are mutually exclusive and that each combination of states of causal factors will determine one particular activity. Each activity, however, can be determined by many different combinations of states, as shown in Fig. 1.1. In the original state space model the levels of causal factors relating to the internal state of the animal were represented as a multidimensional 'physiological space' (McFarland and Sibly 1972, 1975). If, however, several physiological variables could be altered by one particular activity then only one state variable need be represented in 'command space'. Command space is thus viewed as a simplified physiological space. Similarly, causal factors relating to external stimuli could be represented in another multidimensional 'cue space' (McFarland and Sibly 1975). In later versions of the model, however, it has been realized that the separate representation of internal and external stimuli is a complication and that all causal factors, internal or external, may be represented in a single causal factor space and that the motivational state of the animal at any one time may be represented as a point in that space (McFarland and Houston 1981).

Although it is difficult to visualize and represent a multidimensional space, this problem may be overcome by simplification to two or three dimensions. Fig. 1.1, for example, may be viewed as a two-dimensional causal factor space, one vector being the external value of females to a male guppy (cue state), the other being his internal sexual state (command state). We have already seen that some sets of combinations result in the same tendency to court the female and these are said to lie on a 'motivational isocline'. Different combinations of causal factors may therefore give rise to the same activity, although, of course, the same set of causal factors always gives rise to the same activity.

Changes in motivation may be represented as the trajectory of a point in causal factor space; whenever the point is in a particular subset of causal factor space a specific activity will occur. The activity of the animal, however, changes when the trajectory of the point moves from one subset to another. This was demonstrated empirically for the tendency to feed and drink in Barbary doves (Sibly and McCleary 1976) but before discussing that experiment we should briefly consider why the point in causal factor space should move.

As one activity is being performed it may have direct effects on the internal factors relating to that activity; for example, drinking has the effect of increasing blood and cell volume and lowering the osmotic pressure of the blood plasma. Drinking may stop before these internal events occur, however, and we may therefore also postulate changes in causal factors due to the learning of foreseeable consequences of an activity. The occurrence of one activity may also affect internal factors relating to other activities; for example, feeding on dry food may draw water from the blood plasma into the gut, thus changing the tendency to drink. Of course internal factors may also change without being a specific consequence of an ongoing activity.

The cue state may change due to the orientation of the animal changing its perceptual input or because of changes in attention. The cue state may also change when the activity changes the environment. For example, the amount of food remaining will decline during feeding and thus may diminish the cue strength of the food. Finally, the cue state will only be an estimate of the real world, the accuracy of which is limited by the perceptual ability of the animal. As more information is gathered this estimate may change, thus altering the cue state. This final possibility is the basis of assessments.

It is clear that a great many factors may alter the motivation of an animal and, in order to study particular aspects of motivational state, an attempt must be made to keep the other potential variables constant. This was the approach of Sibly and McCleery (1976) when they examined feeding and drinking in doves after deprivation of food and water. The doves were placed in a Skinner box which had two keys; one key gave a food reward when pecked, the other water. However, these rewards were not always available and cues indicated their availability. Although the Skinner box is a very artificial situation, because the amount of food and water were not directly visible to the bird (except for each small reward) no variation in motivation could be attributed to changes in the external environment. Instead it was expected that the animal's feeding and drinking would be determined by its position in causal factor space that was itself determined only by changes in physiological state.

It became evident, however, that not all behavioural change was due to changes in physiological state; some appeared to be due to shifts in attention (McFarland and Lloyd 1973; Sibly and McCleery 1976). In other words, when the physiological state indicated that feeding should be the 'dominant' activity the animal would briefly shift to drinking. If this occurred and the reward apparatus was switched off for a short period the bird would return to drinking when the apparatus was switched on again; that is, it continued with the 'subdominant' activity (McFarland and Lloyd 1973). If the apparatus was switched off for a longer period, however, the bird would be more likely to change back to feeding again, that is, the dominant activity. McFarland and Lloyd (1973) called this phenomenon of brief changes to subdominant activities 'time sharing'. The interruption technique was used to mask behavioural changes due to time sharing and to discover those changes that were due to alteration of the physiological state. Each time the animal changed behaviour the reward mechanism was switched off for a period longer than the duration of brief shifts in behaviour. If the dove resumed the original activity then that was said to be dominant at that time, if it continued the new activity then this new activity was assumed to be dominant. It should be noted that the apparatus was switched off each time the animal pecked the new key and that no reward was given for that peck, preventing any short-term positive feedback effects from occurring. A typical set of results are shown in Fig. 1.7 indicating the points at which behaviour change occurred and which of those changes were due to shifts in dominance. A 'dominance boundary' or 'switching line' may be fitted that results in an almost complete separation of dominant feeding switching points and dominant drinking switching points.

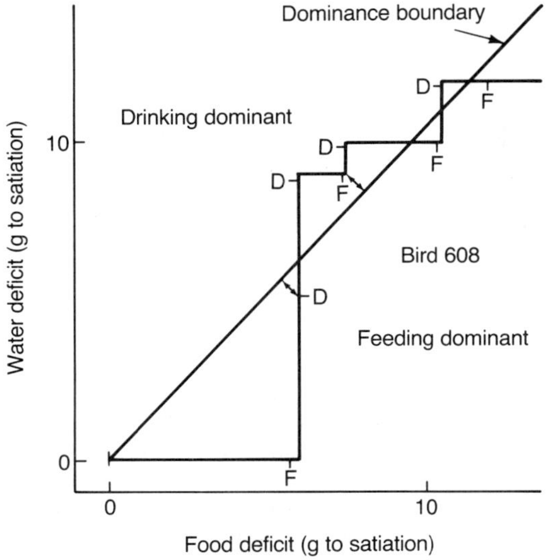

Figure 1.7 The boundary-fitting procedure applied to the data of Sibly and McCleary (1976). F denotes a feeding-dominant point and D a drinking-dominant point according to the criteria in the text.

Not only do doves switch to subdominant activities for short periods during the performance of dominant activities (time sharing) but they may continue with a previously dominant activity after it has become subdominant (Fig. 1.7). For example the causal factors for feeding may be reduced below those for drinking and yet the animal will continue to feed. This phenomenon has been termed 'persistence' and it may be due to short-term positive feedback (McFarland and Sibly 1975). The degree of persistence may be calculated as the 'lock-on' index (Fig. 1.8); the greater the lock-on

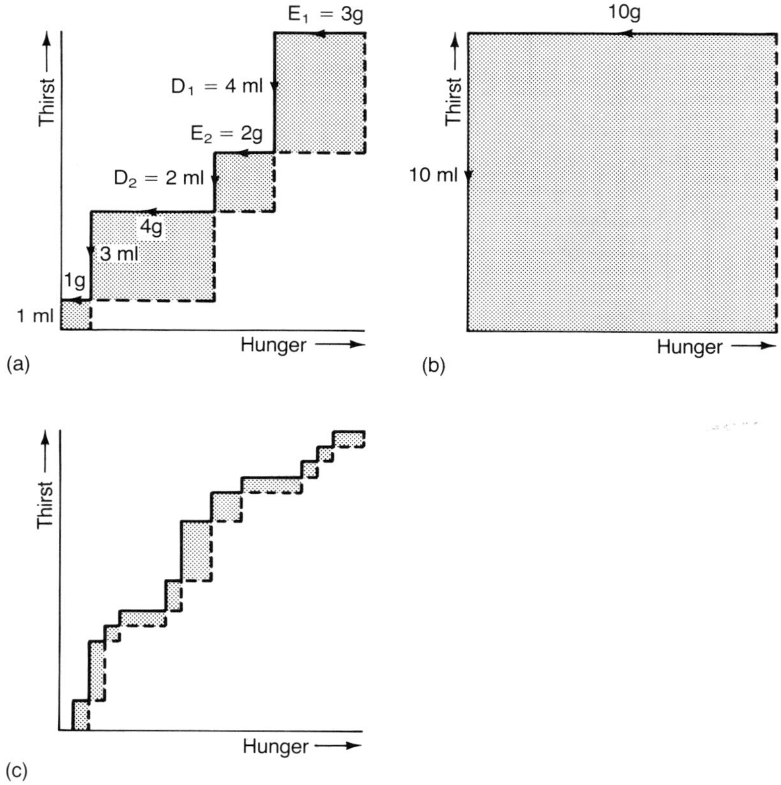

Figure 1.8 Behaviour trajectories followed in satiating hunger and thirst. (a) We multiply each bout of feeding (g) by the following bout of drinking (ml) and add them together in order to calculate a lock-on index. In this case $(3 \times 4) + (2 \times 2) + \ldots$ We arrive at 29 (g × ml). (b) This represents the maximum possible persistence on the part of the animal; it satiates one motivation and then switches. In this case the corresponding calculation gives a value of $10 \times 10 = 100$ (g × ml). Thus the lock-on index of part (a) is actual/maximum = 29/100 = 0.29 and part (b) is 1. (c) In this case the animal alternates frequently between eating and drinking. The grey area is correspondingly small. (From Toates 1980.)

index the greater the persistence. It may be thought that the phenomenon of persistence merely indicates that the motivational system is not perfect. It appears, however, to be an adaptive system that prevents an animal vacillating between two activities. Indeed, as conditions vary so does the lock-on index (Larkin and McFarland 1978). The animal appears to assess the cost of switching behaviour in determining when to switch. If a partition is placed between the keys for feeding and drinking then the lock-on index increases as the partition length increases, i.e. as the cost of changing from one activity to the next increases (Fig. 1.9). This phenomenon has the effect of reducing the time and energy used in switching activities and thus helps to optimize behaviour sequences.

We may infer from these experiments on feeding and drinking in doves that even if the external stimuli are kept constant the physiological state does not completely determine the activities of the animal. Some process or processes influence the motivational state of the animal in such a way as to prevent animals (i) continuing too long with one activity (time sharing) and (ii) switching activities too frequently (persistence). It is clear that in any study of assessments and decisions these phenomena will have to be taken into account.

COMPLEX ASSESSMENTS

During an apparent single assessment procedure, i.e. one leading to a single decision, more than one variable may be assessed. For example, in the case of male *Gammarus* deciding whether or not to take a female into precopula, it has been suggested that the male assesses (i) time to the female's moult (Birkhead and Clarkson 1980), (ii) her size and hence egg number (Ward 1984; Elwood *et al.* 1987), (iii) the number and size of potential competitors (Ward 1983), (iv) the distribution of female states (moult stages and sizes)

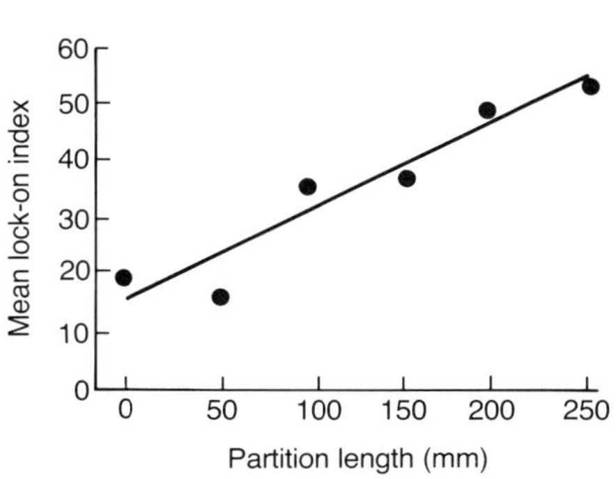

Figure 1.9 Mean lock-on index as a function of partition length. (From Larkin and McFarland 1978.)

in the population (Hunte *et al.* 1985), (v) whether the female is parasitized (Brown and Thompson 1986), (vi) her size relative to the male's size and hence the ability of the male to carry her (Adams and Greenwood 1983) and (vii) the water speed, as this may influence the male's ability to carry the female (Greenwood and Adams 1984; Adams and Greenwood 1987). Even though there is doubt concerning whether or not all of these aspects are actually assessed by the male (e.g. Poulton and Thompson 1987) it is clear that this example of mate choice is more complex than are the simple cases discussed above. Caution should be exercised, however, because in some cases apparently complex assessment procedures may be a series of simple assessments and decisions and in other cases we may be overestimating the assessment abilities of animals.

An example of a series of simple assessments and decisions is found among studies of newt courtship (Verrell 1983). The decision of (i) whether to court quickly using a 'hula' display, (ii) whether to enter into a lengthy amplexus, or (iii) whether to attempt interference is influenced by (a) the presence of another courting male, (b) female receptivity, and (c) whether other non-courting males are in the vicinity. The observed behaviour is complex but may be viewed as a series of yes/no decisions (Fig. 1.10).

An example of overestimation of assessment abilities may be given by a further consideration of the parasitic wasp, *Lariophagus distinguendus.*

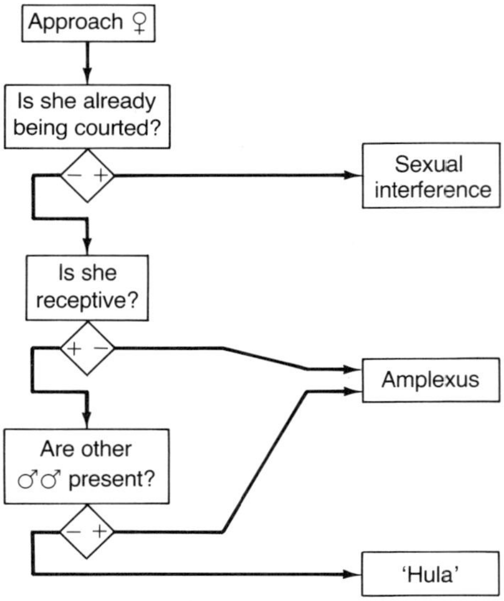

Figure 1.10 The questions posed during a sexual encounter, and the male newt's responses to them. (From Verell 1983.)

Lariophagus does not only have the option of producing a male or female egg but may also reject a host completely (Simbolotti *et al.* 1987). The host lives in cereal grains and not all grains are approached, wasps apparently being able to perceive from a distance whether or not a grain contains a host. Some hosts, however, are approached and then rejected and three main factors appear to be taken into account in this decision: (i) the size of the host as discriminated by tapping the grain with the antennae and also, at a later stage, by contact with the ovipositor (Charnov *et al.* 1981); (ii) the size distribution of the host population; (iii) the density of the host population. It is predicted that hosts larger than a certain size (X_2) should receive a female egg, hosts smaller than a certain size (X_1) should be rejected, and those between X_1 and X_2 should receive a male egg (where $X_2 > X_1$) (Green 1982). A decrease in host density is predicted to shift these threshold values to a lower level whereas an increase in average host size is predicted to shift them to a higher level.

This suggests that a wasp, in order to make perfect decisions, must have a continuously updated representation of the host distribution concerning quality and numbers. The wasp must be able to keep track of its previous decisions concerning oviposition and to recall the intervals between host encounters (to estimate density) (Simbolotti *et al.* 1987). One is immediately led to question whether or not it is reasonable to expect a parasitic wasp to have these abilities or indeed, whether or not it actually conforms to the predictions of optimality. With reference to the latter question it appears that they do not make perfect decisions; some large hosts are rejected and some small hosts receive female eggs (Simbolotti *et al.* 1987). Perhaps the reason for these apparent errors is a lack of ability in assessing the host population.

These considerations led Simbolotti *et al.* (1987) to look for a simpler mechanism that might not be error-proof and yet would approximate optimality. They suggest that one approximate measure of past decisions would be the supply of eggs. A low supply would imply that there has been frequent recent ovipositions because of an abundant supply of good-quality hosts. A large supply would imply a recent paucity of such hosts. Thus egg supply might be used as a 'rule of thumb' concerning past decisions, and thus we do not have to infer great feats of memory to account for the observed behaviour.

These features have been incorporated into a control model of decision-making (Fig. 1.11). The proposed decision-making system has an output in the form of the decision on whether to produce a female egg, a male egg or to reject the host. There are two inputs into the decision-maker, the estimate of host quality and the output of an internal state variable. This internal state variable is essentially a measure of egg supply and this itself has three input variables, the rate of oocyte maturation, feedback from the behavioural

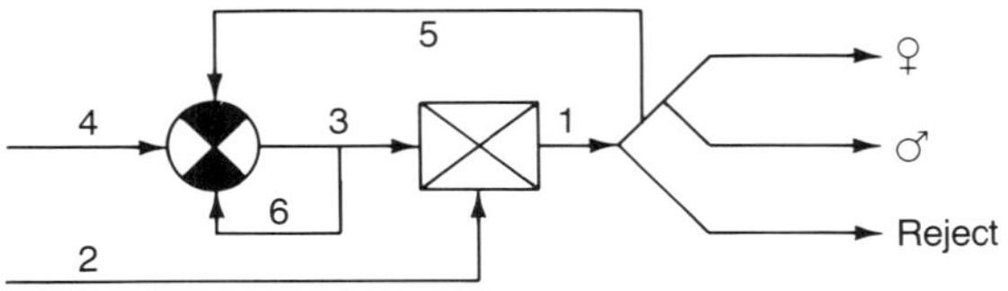

Figure 1.11 A control model for oviposition in a solitary parasite: (1) is the resultant of the interaction of processes triggered by an external stimulus (2), and an internal variable (3), called the 'internal state variable'. Oviposition decisions (female, male, reject) are selected according to the rule: if (1) above threshold b, then oviposit a female egg; if above a and below b, then a male egg, if below a, then reject (b > a). The internal state variable (3) is controlled by the input channels (4), (5) and (6): (4) physiological input (e.g. processes connected with oocyte maturation), (5) behavioural output (i.e. oviposition or no oviposition), (6) physiological imbalance control (e.g. egg resorption). (From Simbolotti *et al.* 1987.)

decision on whether or not to oviposit and an input indicating egg resorption.

The two inputs into the decision-maker may be visualized as a two-dimensional state space (Fig. 1.12) with switching lines delineating sub-spaces for female egg production, male egg production and rejection of the host. It demonstrates how the two main variables interact in the process of decision-making. Computer simulation has tested the predictive value of the

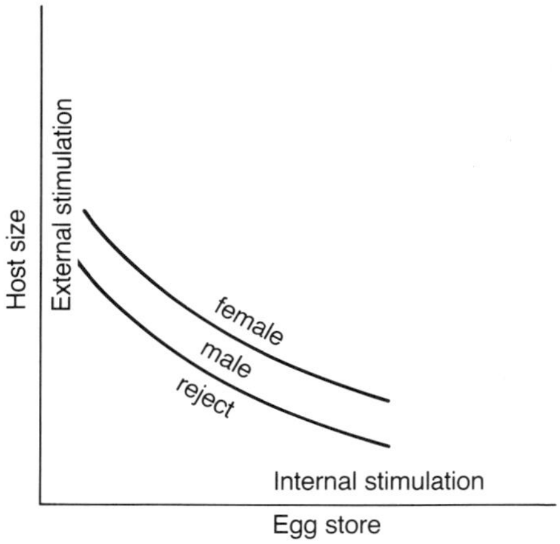

Figure 1.12 The influence of the strength of the external stimulus-situation and the internal state value on oviposition decisions. The curves represent the boundaries (thresholds) for the combination of external and internal factor to result in rejection, oviposition without a fertilization (male) and oviposition with a fertilization (female), respectively. (From Simbolotti *et al.* 1987.)

model and, indeed, it operates well in predicting shifts in the internal state variable in different habitats with a consequent shift in behavioural output (Simbolotti *et al.* 1987). The model is important because it presents a more parsimonious assumption of the animal's assessment abilities and produces clear predictions based upon an easily understood decision-making process. It is this approach that we have attempted to follow in our studies of hermit crabs.

WHY HERMIT CRABS?

Hermit crabs are unusual decapod crustaceans in that they are asymmetrically coiled to fit into gastropod shells (Chapter 2). They are also ideally suited to the study of assessments and decisions. In most animals assessments may only be inferred from the decisions that the animals make, whereas in hermit crabs the assessment procedure itself may be observed. For example, when deciding on whether or not to move into a new shell, a crab will investigate that shell prior to the decision. The sequence of investigatory acts may be recorded and the timing of decisions determined. The shells may be modified to alter the information available to the animal and the effects on the assessment process determined. Furthermore, hermit crabs may fight over the ownership of shells and many of the resource assessment procedures in a fight are similar to those used when empty shells are encountered. Thus motivational models developed to explain the timing of decisions over empty shells may be incorporated into models developed to explain agonistic interactions. These latter cases, however, are more complex as the relative fighting ability of the opponent is also assessed and this feature must also be incorporated into any model of aggressive interactions. Whilst there are a number of excellent studies on assessment strategies during agonistic encounters (e.g. Riechert 1978, 1979, 1984; Austad 1983; Thornhill 1984) we believe that the examination of similar assessment procedures (similar acts and similar information gained) used in agonistic and non-agonistic situations offers unique possibilities for the understanding of animal behaviour.

These concepts of assessments and decisions, as they apply to the behaviour of the hermit crab *Pagurus bernhardus*, will be developed in later stages of this book. However, since readers may have little knowledge of hermit crabs we first review the behavioural ecology of this curious group of animals.

CHAPTER 2

The behavioural ecology of hermit crabs

Hermit crabs first appear in the fossil record in the lower Jurassic, about 190 million years ago. However, these fossils indicate that the animals already possessed an asymmetrical abdomen and that therefore the group arose somewhat earlier (Reese 1983). There are currently more than 700 described species belonging to between 40 and 50 genera (Reese 1969) which are classified within four families in the decapod tribe Anomura. There are two small families, the Coenobitidae including the land hermit crabs and the coconut crab, and the Pylochelidae containing species that generally use holes in rocks, bamboo, sponges or other items rather than gastropod shells to protect their abdomens (see below). There are two large families of marine hermit crabs that are differentiated on the basis of relative cheliped size; in the Paguridae the right cheliped is larger whereas in the Diogenidae either the chelipeds are roughly equal or the left is the larger (Reese 1969). Hazlett (1966a, d, 1968d, e, 1972b) has compared the behaviour of several species and concluded that there are also differences between these taxa with regard to social and agonistic behaviour.

Hermit crabs are found throughout the world in both tropical and temperate waters. They occur from depths of over 800 metres (Ross 1967) to the intertidal zone, particularly where gastropod shells are available (Fotheringham 1980). Although there do not appear to be any freshwater species, some species enter estuaries for at least part of the year (Roberts 1971a; Rebach 1978). Some land hermit crabs (Coenobitidae) have become so well adapted to a terrestrial existence that they need to return to the sea only to reproduce (Carpenter and Logan 1945; Matthews 1956b; Bliss 1968; Gruber and Shoup 1969; De Wilde 1973). They can live up to several miles inland and prefer dry areas, even living on mountains up to 900 metres high (De Wilde 1973).

The use of gastropod shells has been an important factor in allowing tolerance of such broad environmental conditions. In addition to protecting crabs from predation the possession of a shell greatly increases the animal's tolerance to desiccation (Young 1978; Taylor 1981) and extremes of salinity

(Shumway 1978). The possession of shells was probably the decisive factor in allowing the Coenobitidae to become so strongly terrestrial (Radinowsky and Henderson 1974; McMahon and Burggren 1979). These animals hold water in their shells which they regulate to the salinity of sea water using fresh water and water in their food (De Wilde 1973). Adults of the land coconut crab (*Birgus latro*) do not carry shells (although juveniles do), and have compensated for this through behavioural modifications including nocturnality and inhabiting burrows, as well as possessing a secondarily toughened abdomen (Reese 1968b; Helfman 1977a, b).

Some species have adopted objects other than typical gastropod shells to protect their soft abdomens. Although the abdomens of most hermit crabs are asymmetrical and adapted for the use of dextrally curved gastropod shells, crabs will utilize left-handed shells or highly modified shells, such as those of cones, if they are given no choice (MacKay 1945b). Large *Coenobita* in Bermuda utilize large fossil shells where there are no other shells of a suitable size available (Haas 1950). Some symbionts (especially anemones and hydrocorals) modify and extend the crab's shell to a great extent, sometimes to the extreme that the shell itself disappears entirely (Ross 1967; Jensen 1970; Ball and Haig 1974; Slack-Smith 1982). Other crabs, many belonging to the family Pylochelidae, have become secondarily symmetrical and use sponges, scaphopod shells, pieces of bamboo or coral, serpulid or other worm or vermetid gastropod tubes, or cavities in corals, sponges or stones (Hazlett 1966a, 1969c; Provenzano 1968; Markham 1977; Schumacher 1977; Caine 1980; Patton and Robertson 1980). Many of the species inhabiting attached tubes or cavities are filter feeders and rarely, if ever, leave their cavities (Markham 1977; Schumacher 1977; Caine 1980). One species, *Paguritta harmsi*, living in coral heads, lives in aggregations, with males and females living in adjacent tubes (Patton and Robertson 1980). In another species, *Calcinus verrilli*, some individuals are found living in attached vermetid shells whereas others live in unattached shells (Markham 1977).

Many species of hermit crabs have diurnal or tidal periods of activity. Clustering during low tide and dispersing to feed when the tide returns occurs in several species (Ball 1968; Snyder-Conn 1980; Taylor 1981) and is thought to improve survivorship by increasing resistance to cold, desiccation and physical displacement (Snyder-Conn 1981). In at least one species these rhythms persist in conditions of constant light and water, indicating endogenous control (Imafuku 1981). Clustering during the day and dispersal at night occur in semi-terrestrial *Coenobita* (Ball 1972). Terrestrial and semi-terrestrial species of *Coenobita* tend to be nocturnal as adults, but juveniles vary and their activity patterns are strongly affected by temperature and humidity (Niggemann 1968; Ball 1972; De Wilde 1973). Marine crabs tend to exhibit activity peaks, but these may be either diurnal or nocturnal depending upon the species (Hazlett 1966d, 1981b; Ball 1968; Mitchell 1973; Ameyaw-Akumfi 1975; Stachowitsch 1979) and may or may not persist when the animals are held in constant environmental conditions (Ball 1968; Mitchell 1973; Imafuku 1981). Some intertidal species migrate offshore in the winter to avoid severe conditions in the intertidal zone (Pike and Williamson 1959; Samuelson 1970; Rebach 1974, 1981; Fotheringham 1975), and some of them apparently spend much of the cold weather inactive and buried in the sand (Rebach 1974). Other species commonly bury themselves in the sand to avoid predators (Wicksten 1979).

Physiological tolerance to desiccation and osmotic stress is generally greater in inshore animals than in those living offshore, with both different species and different populations of a single species showing varying degrees of tolerance (Bollay 1964; Davenport 1972; Young 1978, 1979b, c, 1980). Intertidal animals appear to have adapted both behaviourally and physiologically to a more physically stressful environment, but one that offers some degree of protection from predation and is relatively rich in nutrients (Reese 1969; Sabourin and Stickle 1980; Bertness 1981f; Burggren and McMahon 1981). Size also seems to affect stress tolerance, with small *Coenobita compressus* more sensitive to desiccation than large animals (Ball 1972). When subjected to low salinity levels intertidal *Pagurus bernhardus* retreat into their shells and decrease their oxygen consumption, and this behaviour allows the animals to survive short periods of extreme osmotic stress (Davenport *et al.* 1980).

Although the use of shells offers some protection from many predators, hermit crabs are not immune from predation. Many of their predators have adaptations for feeding on gastropods and use the same means to take hermit crabs. Several species of crabs and lobsters crack mollusc shells open to extract the contents, either snail or hermit crab (Wilson 1949; Shoup 1968; Berman *et al.* 1973; McLean and Mariscal 1973; Zipser and Vermeij 1978; Bertness 1981f, 1982; McLean 1983). Fish, such as cod and dogfish, take crabs by either crushing the shell (Jackson 1913; Bertness 1981f, 1982) or biting off the anterior portion of the crab (Brightwell 1952). Different gastropod species or populations have shells adapted to minimize predation (Bertness and Cunningham 1981; Bertness *et al* 1981; Currey and Hughes 1982), and hermit crabs use different strengths of shell in areas subject to varying degrees of predation (Bertness 1981f, 1982). Crabs inhabiting shells that are smaller than preferred have been demonstrated to be more vulnerable to predation than are crabs in adequately sized shells (Vance 1972b).

SYMBIOTIC ASSOCIATIONS

Protection from predation by some species has been achieved through symbiotic partnerships between hermit crabs and other organisms, especially coelenterates. Crabs with symbiotic anemones show reduced susceptibility to attacks by crabs (Shoup 1968; McLean and Mariscal 1973) and octopus (Ross 1971; Balasch and Mengual 1974; McLean 1983). Hermit crabs with symbiotic hydroid colonies on their shells are also less vulnerable to certain types of predation (Berman *et al*. 1973; Grant and Pontier 1973; Brooks and Mariscal 1985). The importance of predation in maintaining the association between hermit crabs and anemones was demonstrated by Balasch and Mengual (1974) and by Ross and von Boletzky (1979). The anemone *Calliactis parasiticus* is a common associate of *Dardanus arrosor*, with the crabs actively acquiring anemones if their own anemone is removed. In captivity this activity eventually wanes, but reappears suddenly if an octopus is introduced into the tank with the crabs (Balasch and Mengual 1974) or water from a tank containing an octopus is introduced (Ross and von Boletzky 1979). The possession of anemones is influenced by the dominance status of the crab, with dominant animals maintaining anemones in contests (Mainardi and Rossi 1969, 1971).

This association between crab and anemone appears to be mutually beneficial, with both partners exhibiting behaviour patterns leading to its establishment and maintenance (Cowles 1919, 1920; Schiejsma 1935; Ross and Sutton 1961a, b, 1968; Fox 1965; Ross 1967, 1970, 1975; Cutress and Ross 1969; Ross and Wada 1975). Anemones and other animals that associate with hermit crabs gain a site of attachment (the shell), mobility and some protection from their own predators (Conover 1975; Stachowitsch 1979; Taylor 1979). Some also obtain food directly or indirectly from the crab (Jackson 1913; Orton 1927; Brightwell 1951; Fox 1965; Jensen 1970; Fotheringham 1976b).

The epibiont communities on shells can at times be quite large and complex, especially in regions of soft sediment where there are few other sites for attachment of organisms (Ball and Haig 1972, 1974; Jensen and Bender 1973; Fotheringham 1976b; Stachowitsch 1977, 1980; Markham 1978; Telford and Daxboeck 1978; Karlson and Coriolou 1982). Protozoa, sponges, coelenterates, flatworms, polychaetes, bryozoans, copepods, barnacles, amphipods, isopods, decapods, gastropods, lamellibranchs and tunicates have all been found associated with *Pagurus bernhardus* shells (Jensen and Bender 1973). The relationship of many of the organisms with the crab is unknown. Some, especially the coelenterates, are beneficial, providing protection and even enlarging or replacing the shell (Selbie 1921; Wilson 1949; Fox 1965; Jensen 1975; Hazlett 1981a; Slack-Smith 1982), or providing sustenance for the crab which may eat food captured by the anemone or even the anemone's eggs (Brightwell 1951, 1952). Others are probably harmful; many species will consume hermit crab eggs (Christensen 1967; Fotheringham 1976b) and some may reduce the space within the shell available for carrying eggs (Fotheringham 1976b). Others eat food that the crab has obtained (Jackson

1913; Brightwell 1951), and some may add significantly to the weight of the shell, thus making it less suitable for the crab (Conover 1976). Different species of crab show preferences for shells with particular symbionts; some species prefer epibionts which others avoid (Grant and Ulmer 1974; Wright 1973; Conover 1976; Mercando and Lytle 1980; McLean 1983). The presence of some symbionts, especially barnacles and gastropods, will cause crabs to leave the shell when possible (Brightwell 1952; Conover 1976), whereas other symbionts (primarily anemones and decapods) will be transferred when the crab changes to a new shell (Telford and Daxboeck 1978; Hazlett 1981a). Symbionts also affect both inter- and intraspecific interactions among hermit crabs by affecting the desirability of shells (Mainardi and Rossi 1969; Grant and Ulmer 1974; Wright 1973; Conover 1976, 1979; Karlson and Coriolou 1982).

FEEDING

Hermit crabs are omnivorous with three basic modes of feeding: detritivory, filter feeding and macrophagus scavenging (Samuelson 1970; Caine 1975; Kunze and Anderson 1979; Schembri 1982). Probably the most generalized are the bottom-scavenging detritivores which sift small organic particles from the substrate using their chelae and mouthparts. Many of the crabs inhabiting attached tubes have become specialized to varying degrees for filter feeding, using their antennae (Schumacher 1977; Caine 1980) or forcing a current past their mouth parts and thoracic appendages using their antennules (Snow 1973; Markham 1977). Some species (perhaps many) combine these two modes of feeding opportunistically (Boltt 1961; Greenwood 1972). For example, *Pagurus bernhardus* is primarily a detritivore but also feeds by filtering small plants and animals from suspension using its antennae (Jackson 1913; Orton 1927; Gerlach *et al.* 1976), and lies on its back and waves its legs to obtain detritus from the surface foam (Neil pers. obs.), a behaviour also reported in *P. longicarpus* (Scully 1978). In contrast, *Discorsopagurus schmitti*, a crab living in attached worm tubes, feeds primarily by filtration using its antennae, but will also consume any detritus settling near its tube (Caine 1980). The third mode of feeding exhibited by hermit crabs is macrophagous scavenging. Many marine species will eat large pieces of detritus when they are encountered (Orton 1927; Randall 1964), but most species have mouthparts and digestive tracts adapted for smaller particles (Roberts 1968; Caine 1975, 1976; Kunze and Anderson 1979). The Coenobitidae, including the coconut crab, appear adapted primarily to macrophagous scavenging, eating most types of animal and vegetable organic matter (Reyne 1939; Ball 1972; De Wilde 1973).

REPRODUCTION

In many hermit crab species males are larger than females (Ball 1972; De Wilde 1973; Ameyaw-Akumfi 1975; Fotheringham 1975), and the suggestion that some species may be protogynous sequential hermaphrodites has been made (Wenner 1972). However, it is probable that for most species males simply grow faster than females (Markham 1968), possibly due to the inhibiting effect of egg-production on female growth (Bertness 1981c, g). Mating behaviour is often complex (Hazlett 1966d, 1968e, 1970c, 1972b; Ameyaw-Akumfi 1975), but can be simple, for example in the coconut crab (Helfman 1977b). Copulation does not appear to be tied to moulting by the female (De Wilde 1973; Helfman 1977b) and opinion is divided upon whether fertilization is internal (Kamalaveni 1949; Ameyaw-Akumfi 1975) or external (Matthews 1956b). Male *Pagurus geminus* discriminate between receptive females and other males by a water-borne chemical detected by the antennules (Imafuku 1986).

The number of eggs produced is directly related body size, with females often starting to breed when they are quite small (Yamaguchi 1938; Bollay 1964; Samuelson 1970; Fotheringham 1975, 1976c). As with most aspects of hermit crab biology, the shell has a strong effect on reproduction. Animals with adequate or large shells may grow, but if their shells are small they are more likely to put energy into reproduction (Bertness 1981c, g). Other studies, however, have suggested that clutch size is decreased if the shell is too small (Fotheringham 1975, 1976c; Bach *et al.* 1976).

The female carries the eggs attached to her pleopods (biramous abdominal appendages), usually until they hatch, and then releases the larvae into the sea by moving rapidly in and out of her shell (Coffin 1960). In terrestrial species release of larvae involves the adults migrating up to several miles to the shore (Yamaguchi 1938; Carpenter and Logan 1945; De Wilde 1973). Reproduction is matched to local environmental conditions (Reese 1968a) and may or may not be seasonal (Pike and Williamson 1959; Coffin 1960; De Wilde 1973; Ameyaw-Akumfi 1975; Fotheringham 1975; Fotheringham and Bagnall 1976; Bertness 1981g; Hazlett 1983b). Reproductive patterns have also been shown to be affected by interspecific competition (Ameyaw-Akumfi 1975; Fotheringham 1980; Bertness 1981d, f, g).

Following their release by the female the larval crabs spend most of their developmental stages in the plankton, where they may stay for from three to eight or more weeks (Pike and Williamson 1959; Coffin 1960; De Wilde 1973). Several studies have been conducted to elucidate the larval stages of various species of hermit crabs (Thompson 1903; McDonald *et al.* 1957; Provenzano and Rice 1966; Reese and Kinzie 1968; Roberts 1970, 1971a, b; Provenzano 1978). It is only in the final larval moults that the crabs develop an asymmetrical body (Agassiz 1875; Roberts 1970). The glaucothoe (young crab stage) first enter shells, and they are able to enter shells without prior experience (Agassiz 1875, 1876; Jackson 1913; Reese 1962a), though this depends on the conditions under which they have been reared (Hazlett and Provenzano 1965; Hazlett 1971a). Sexual maturity in different species may be reached in as little as four months (Fotheringham and Bagnall 1976) or as much as one year (Jackson 1913). Species also vary in the time taken to reach maximum size: e.g. 12 months in *P. longicarpus* but 3 years or more in *P. pollicarus* (Fotheringham and Bagnall 1976). Some species, e.g. *P. bernhardus*, continue to grow for many years (Jackson 1913).

SHELL ACQUISITION

A crab can obtain a gastropod shell in one of three ways: it can locate an empty shell, obtain a shell from another hermit crab, or obtain a shell directly from a gastropod. In practice it appears that crabs only rarely kill snails to obtain their shells (Rutherford 1977), though crabs have been seen to eat injured gastropods and then use their shells (Agassiz 1875, 1876; Greenwood 1972; Imafuku 1983a), and sick or injured gastropods may be extracted from their shells and discarded by crabs prior to using their shells (Brightwell 1951). It appears to be common for crabs to be attracted to sites at which predators attack gastropods and for the crabs to appropriate the shells after the predators have consumed the occupants (Brightwell 1952; McLean 1974, 1983; Hazlett 1980a, 1982a; Hazlett and Herrnkind 1980; Rittschof 1980a, b; Wilber and Herrnkind 1982, 1984). McLean (1974) reports large numbers of crabs attracted to sites of gastropod predation, with the crabs forming a hierarchy in which the most dominant animal gained first access to a shell and, if rejected by the dominant crab, the shell was passed on down the hierarchy. Once a crab decided to occupy the shell it vacated a shell which then became available to another crab and so on. This process has been investigated and termed 'the vacancy chain process' in which resources are passed from one individual to another (Chase *et al.* 1988).

Attraction to predation sites is initially by orientation to chemicals released by injured gastropods and secondarily by vision (McLean 1974, 1983; Hazlett 1980a, 1982a; Hazlett and Herrnkind 1980; Rittschof 1980a, b; Wilber and Herrnkind 1984). This chemical attraction can be very specific, with hermit crabs only attracted to particular species of gastropods that have shells of an appropriate size for them (McLean 1983). Chemical cues also appear to be used in identifying shells. Mesce (1982) has shown that crabs prefer artificial shells that bear calcium to those that do not, and thus they are presumably using the presence of this element as a cue that an object being investigated is in fact a gastropod shell.

The availability of empty shells varies greatly for different populations of crabs. The behaviour of crabs towards shells, and the degree to which the crabs are satisfied with their shells, varies with shell availability (Scully 1979, 1983b). Most workers have found relatively few unoccupied shells where hermit crabs are abundant (Haas 1950; Provenzano 1960; Hazlett 1970b; Ball 1972; Childress 1972; Vance 1972a; Fotheringham 1976b; Kellogg 1976; Spight 1977; Elwood *et al.* 1979; Scully 1979; Imafuku 1983a, b). In a single area the availability of shells may be greater for one size range of crabs than for another (Berman *et al.* 1973; Young 1979b). Shell availability can also fluctuate rapidly, as shells are buried in soft sediment by natural wave action, reducing their availability to hermit crabs (Conover 1975; Stachowitsch 1977). There is considerable evidence that low shell availability limits hermit crab numbers (Provenzano 1960; Vance 1972a; Berman *et al.* 1973; Radinowsky and Henderson 1974; Fotheringham 1976b; Kellogg 1976; Spight 1977; Scully 1979; Wang and Jillson 1979). This limitation can be general, or may operate during only one phase of the life cycle, either for small (Wang and Jillson 1979) or large crabs (Berman *et al.* 1973; Kellogg 1976). Several field studies have shown that crabs tend not to be housed in optimal shells, thus supporting the idea that shells are often a limiting factor (Childress 1972; Abrams 1978; Imafuku 1984). Imafuku (1983b) reports several observations of two hermit crabs occupying a single shell, and this appears to be a consequence of low shell availability (the crabs involved were not reproductive pairs). Further evidence that hermit crab numbers are limited by shell availability comes from recent work showing that crabs rapidly start to use shells of introduced species (Wicksten 1977; Blackstone and Joslyn 1984).

WHAT IS AN 'OPTIMAL SHELL'?

Bertness (1981b, f) has shown that different aspects of a hermit crab's fitness may be optimized by different species of shells; the type and size of shell an animal possesses influences its survival, growth and reproduction (Bertness 1980, 1981c, g). Crabs in shells too large for them grow while those in small shells reproduce, which results in crabs with a poor supply of shells being smaller and reproducing sooner and more frequently than do crabs with a good supply of shells (Bertness 1981e; Scully 1983b). However, since clutch size in hermit crabs is closely related to crab size the diminished size of the crabs may result in a decreased lifetime reproductive success compared to crabs that have an adequate supply of shells (Fotheringham 1976a, c, 1980), and Fotheringham (1976b) has shown a decrease in egg numbers in crabs in inadequate shells. Crabs in optimally sized shells grow faster than do those in shells too small for them (Markham 1968; Fotheringham 1976a), and they are also less susceptible to predation (Vance 1972a) and physiological stress (Taylor 1981). Even a crab's shape may be influenced by the shells it inhabits.

Crabs inhabiting shells with long narrow apertures exhibit a greater degree of dorso-ventral flattening than do conspecifics in shells with rounder apertures (McLaughlin and Bailey-Brock 1975), and *P. bernhardus* inhabiting scaphopod shells have flatter bodies and more symmetrical abdomens and chelipeds than do crabs in typical gastropod shells (Selbie 1921). Blackstone (1985) has shown that the size and shape of the shells a crab uses have an effect on the crab's ultimate size and even on its growth rate and the relative dimensions of its chelipeds.

Different species and different sizes of crabs within a species show preferences for different species of shells (MacKay 1945b; Bollay 1964; Orians and King 1964; Grant and Ulmer 1974; Mitchell 1975; Sripathi *et al.* 1977; Elwood *et al.* 1979; Bertness 1980). Individual crabs prefer particular sizes of shells, with shell weight, volume, length and aperture size all being implicated as features influencing this preference (Reese 1962a; Grant 1963; Volker 1967; Childress 1972; Kuris and Brody 1976; Mitchell 1976; Conover 1978; Gilchrist 1984). Even the colour of shells can be important, with crabs tending to choose shells which match their background in preference to those which contrast (Partridge 1980). Symbionts also influence shell preferences, with some crab species preferring shells with symbionts and some preferring those without (Jensen 1970; Lindsey and Grant 1970; Grant and Pontier 1973; Grant and Ulmer 1974; Conover 1976; Mercando and Lytle 1980). Crabs have also been shown to prefer shells that have been previously occupied by themselves or by conspecifics (Grant 1963; Abrams 1978). In the case of *Coenobita* this preference is due to previous occupants eroding the columella which increases the effective size and decreases the weight of the shell (Kinosita and Okajima 1968; Ball 1972; Abrams 1978).

It is clear from the discussion above that a variety of benefits accrue to hermit crabs by inhabiting shells, but there are also costs. The energetic cost of carrying a shell has been calculated for the land hermit crab, *Coenobita compressus* (Herreid and Full 1986a). Oxygen consumption is about 50% greater in crabs carrying shells than for naked crabs over a wide range of velocities (Fig. 2.1) and the oxygen consumption increases as the load increases. In aquatic hermit crabs the effect of the weight of the shell may not be as marked due to the greater density of the medium. Movement of the crab, however, might be impeded because of increased friction between a large shell and the water. It is thus important that a crab should obtain a shell large enough to gain the benefits discussed above and yet to minimize the energy costs of carrying the shell. In *Coenobita* a preference is shown for light shells of the genus *Nerita*, even though this type confers less protection than many others (Abrams 1978; Bertness 1981c; Herreid and Full 1986b).

Different species of shell may also confer different fitness on the occupants due to differences in mating success. Males inhabiting some species of shell obtain more matings than do males in other species and experiments in which

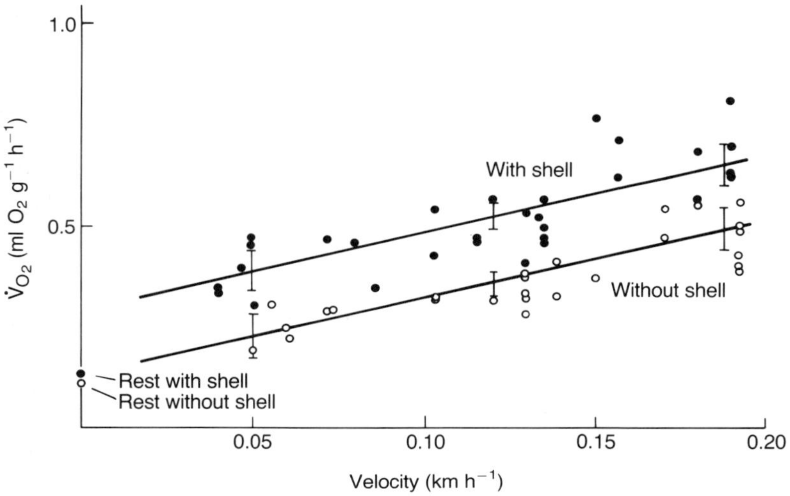

Figure 2.1 Oxygen consumption ($\dot{V}_{O_2}$) of hermit crabs (*Coenobita compressus*) running at different velocities (V). The solid circles represent crabs carrying snail shells as they walked on a treadmill and the open circles are crabs without shells. The lines through the points are least-squares regression lines: with shell, $\dot{V}_{O_2} = 0.29 + 1.98\,V$; without shell, $\dot{V}_{O_2} = 0.12 + 2.00\,V$. The error bars are 95% confidence intervals. (From Herreid and Full 1986a.)

males have been given alternative shells have shown this effect to be mediated by the shell rather than the competitive ability of the males (Hazlett and Baron 1989). This appears to be because of two main factors; firstly, males in some species 'handled' females poorly during precopulatory behaviour and often dropped them, and secondly, females showed a greater responsiveness to males in particular species of shells.

INTERSPECIFIC COMPETITION

Although most competitors for shells that a crab encounters are likely to be conspecifics (Bertness 1981d; Abrams 1982b), there can also be interspecific competition over shells between sympatric species. Two recent reviews of hermit crab research have emphasized the importance of competition in the evolution and behaviour of hermit crabs (Hazlett 1981a; Scully 1983a). Several studies have shown that interspecific competition has an adverse effect on the mean quality of shells utilized by crabs, and hence on their survival and reproduction (Hazlett 1970b; Bach *et al.* 1976; Abrams 1981a; Bertness 1981a; Imafuku 1984). Other studies have shown that different crab species partition the shell resource so that there is now little direct competition for shells, though the fact of this partitioning indicates that competition may have been an important force in the past (Provenzano 1960; Lindsey and Grant 1970; Mitchell 1975; Bach *et al.* 1976; Kellogg 1977; Abrams 1980, 1981b, c, 1982a; Hazlett 1980a). Studies of competition between hermit crab species have shown that some species tend to dominate others, and that the effects of such competition often lead to lower shell quality in the subordinate species (Reese 1961; Bach *et al.* 1976; Abrams 1981c, 1982a; Bertness 1981a). Interspecific competition is not limited to competition with other hermit crab species; crabs also compete for empty shells with fish (gobies and blennies) and octopus in some areas (Mather 1982; McLean 1983).

To summarize, shells are extremely important resources to hermit crabs. Having shells of the correct species and size can increase survivorship and reproduction. Competition over shells is often intense, with the availability of shells often serving to limit population size. Interspecific competition influences the distribution of different species, and can also result in direct contests between individuals of different species over shells. However, in general it appears that intraspecific competition for shells is more important than interspecific.

CHAPTER 3

Pagurus bernhardus in Northern Ireland

This chapter will present information on the biology and shell preferences of the common European hermit crab, *Pagurus bernhardus*. As the name implies this crab is widely distributed in the seas and shores around Europe and is readily collected from our study site, the littoral zone on the eastern side of the Ards Peninsula, County Down, Northern Ireland (Fig. 3.1). Detailed anatomical and taxonomic descriptions of this species may be found elsewhere (Jackson 1913; Selbie 1921; Lancaster 1988). Suffice it to say that *P. bernhardus* is a typical pagurid with the right cheliped larger than the left. Apart from the chelipeds *Pagurus* has two pairs of typical walking legs (pereiopods) and two pairs that are reduced in size and normally held within the shell. Together, these appendages constitute the five pairs of thoracic appendages that give the name to the order, Decapoda, to which hermit crabs belong (Chapter 2). The genital openings are on the coxal segments of the 5th pereiopods in the male and on the 3rd pereiopods in the female. The normal decapod pattern of five pairs of abdominal appendages or pleopods is modified in *Pagurus* in that they are missing on the right side which is held against the collumela of the shell (Fig. 3.2). In the male small biramous pleopods 1, 3 and 5 are present on the left but one of the ramii on each is reduced. The female has four pleopods on the left (1, 2, 3 and 5) and these are larger than in the male with the ramii more equal in size; they are used by the female to carry eggs. The sixth pair of abdominal appendages, the uropods, are asymmetrically developed and function to anchor the animal within the shell (Elmhirst 1947).

The crabs occur primarily in areas where medium to large rocks are interspersed with sandy areas, and are particularly abundant at low tide in tide pools with sandy bottoms. Crabs often occur under clumps of seaweed (especially serrated wrack, *Fucus serrata*) growing on the rocks and hanging over a pool of water.

The abundance of crabs fluctuates greatly. At times finding any crabs is difficult, whereas at other times several can be found under most clumps of seaweed. In some circumstances the crabs can become extremely aggregated

Figure 3.1 The collection site at Ballywalter on the east coast of N. Ireland.

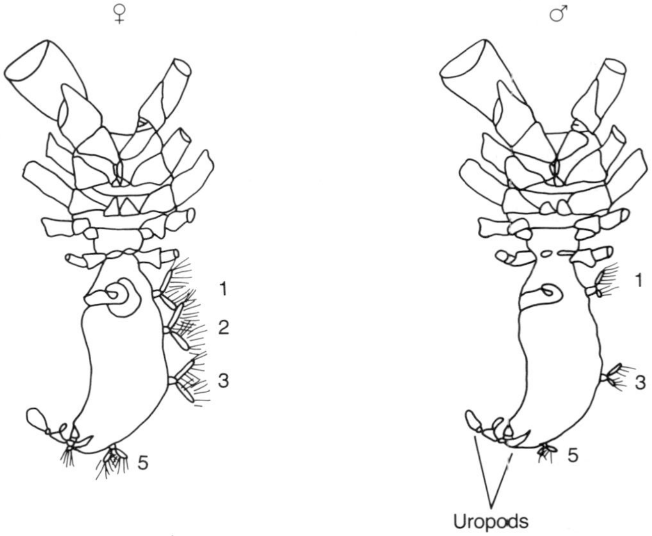

Figure 3.2 *Pagurus bernhardus* showing ventral view of abdomen. The female on the left possesses four filamentous pleopods whereas the male, on the right, has only three pleopods which are reduced (pleopods are numbered).

in their distribution with several dozen occurring within a square metre, this being particularly the case in very cold conditions. Wind and rainfall also appear to influence crab distribution, with animals being more abundant intertidally if the weather is calm with little rain than in periods of strong wind and heavy rain; however, there do not seem to be large seasonal changes in the abundance of crabs. There has been no noticeable change in abundance in the eleven years during which this population has been studied. Although there are inevitably some casualties among the animals brought into the laboratory for study, the vast majority are returned unharmed to the shore within two weeks of collection.

Three parasites regularly occur in our population, the cirriped *Clistosaccus paguri* (Rhizocephala) and the isopods *Athelges paguri* and *Pseudione hyndmanii* both in the family Bopyridae. The cirriped is an invasive parasite which sends 'roots' deep into the abdomen of the crab leaving only a small featureless mass visible from the exterior. We had first identified it as *Peltogaster paguri* but Lancaster (1988) points out that *Peltogaster* is a vivid red whereas *Clistosaccus* is white. *Peltogaster* is commonly referred to in the literature whereas *Clistosaccus* is rarely mentioned (Lancaster 1988), although the latter appears to be the only parasitic cirriped at our study site. *Athelges* and *Pseudione*, on the other hand, are ectoparasites exhibiting fairly standard isopod features except that they too are asymmetrically developed (see Markham 1978 for a review of similar parasitic isopods from the Northwestern Atlantic Ocean). *Athelges* is found on the abdomen whereas *Pseudione* is found within the branchial cavity. The incidence of parasitism is low, probably less than 5% of the population.

More common, and affecting roughly 10% of the population, is the occurrence of missing or regenerating limbs. Limbs can be lost, or damaged and later autotomized, in either predation attempts or agonistic encounters. The relative contributions of these two factors are unknown, but limb loss does occasionally occur in the holding tanks in the laboratory that contain only *P. bernhardus*, indicating that intraspecific conflict can lead to such major injuries. Similar injuries have been noted in other species (Hazlett 1966b, d). Detachment of a small part of one walking leg was observed during one agonistic encounter, and in another encounter in which a recently moulted crab was attacked it lost both chelipeds and a walking leg, but these were the only observations of such damage in hundreds of hours of observation of agonistic behaviour. The limbs most commonly missing in crabs are one or both of the chelipeds, but crabs may also lose one or more walking legs, sometimes in addition to the loss of chelipeds. Limbs lost begin to regenerate at the subsequent moult, and regenerating limbs (identifiable by their small size and slightly different colour) are common.

Pagurus bernhardus is the only species of hermit crab to be found on the shore of our study site, although other species may be found sublittorally

(Pike and Williamson 1959). The vast majority (>97%) of these crabs on the shore occupy five species of gastropod shells, these being *Littorina obtusata (littoralis)*, *Gibbula cineraria*, *G. umbilicalis*, *Nucella (Thais) lapillus* and *L. littorea*. A few small crabs (< 3%) occupy *L. saxatilis* and *Nassarius reticulatus* shells and some large crabs may be found in *Buccinum undatum* shells. The largest crabs, however, are restricted to sublittoral zones and inhabit both *B. undatum* and *Neptuna antiqua* shells.

Several surveys of shells occupied by crabs and those available for occupancy have been conducted by searching at low tide for all shells that are not occupied by molluscs. The most common shell on our shore is *L. obtusata* and about 80% of the crabs occur in this species (Table 3.1). Different sizes of crabs, however, occupy different species of shell (Fig. 3.3) with most crabs less than 0.5 g inhabiting either *L. obtusata* or *Gibbula* species (for most of our studies the two *Gibbula* species are not discriminated as they vary mainly in their markings, the size and shape being similar). Most crabs larger than 0.5 g inhabit *N. lapillus* or *L. littorea*. Thus we may regard crabs as moving through a succession of species in their lifetime, first either *L. obtusata* or *Gibbula* shells, then to either *N. lapillus* or *L. littorea* and finally to *Buccinum undatum* or *Neptuna antiqua*, which occur sublittorally.

REPRODUCTION

A recent study examined the mating behaviour of *P. bernhardus* and the fitness consequences of occupying particular shells (Elwood and Marks unpublished). Crabs in *Littorina obtusata* shells were collected at two-week intervals from late October to early January. Equal numbers of both males and females were then allocated to *L. obtusata* and *G. cineraria* shells, placed in groups of 20 animals and observed for a one-hour period. Six such groups were observed for each two-week interval. There was a significant increase in interactions initiated by males towards females, starting in late November and reaching a peak in late December. At the same time female to male interactions declined. This period coincides with the appearance of females

Table 3.1. Combined data of several shore surveys. Data concerning shells still occupied by gastropods are excluded

	Total	*Vacant*	*Occupied by crabs*
L. obtusata	3091	185	2906 (94%)
Gibbula sp.	463	219	244 (53%)
N. lapillus	241	21	220 (91%)
L. littorea	146	9	137 (94%)

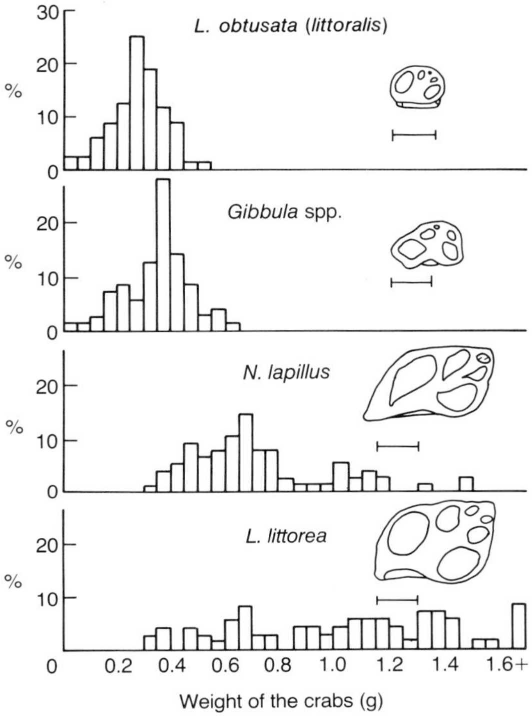

Figure 3.3 The distribution of different sized crabs is plotted for each of the four main species of shell. The cross-sectional shape of each species, showing the internal configuration, is also given. A 1 cm scale is provided to allow comparison. (From Elwood *et al.* 1979.)

carrying eggs in this population (see below). Some of these interactions were sexual in nature but the majority seemed to be either shell fights or agonistic displays. Sexual interactions were defined as being those that had precopulatory patterns (described below). The percentage of interactions classed as sexual rose from 1% in late October to 19% in late December (Fig. 3.4). Of the 103 sexual interactions 15 resulted in mating.

Prior to mating the male grasped the rim of the female's shell with his minor cheliped, with the female facing away from the male. This phase lasted 10–25 minutes during which time the male defended the female from approaches of other animals by displays of the major cheliped. The female was then turned to face the male while he continued to hold the rim of her shell. The male often held his chelipeds over the aperture of the female's shell in a characteristic posture (Fig. 3.5). He then began to pull at the chelipeds and ambulatory legs of the female and held the female's shell in place by

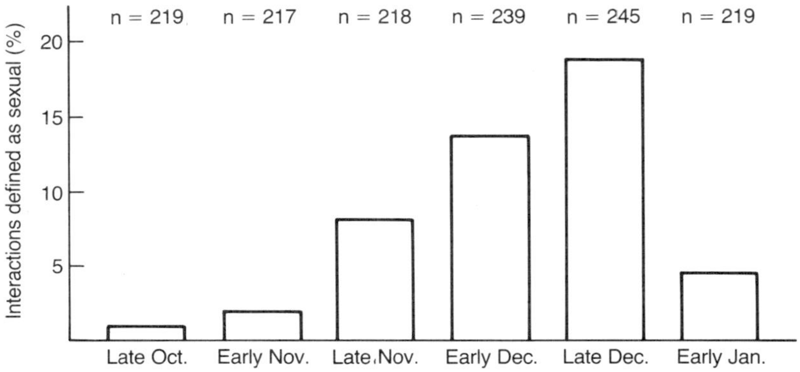

Figure 3.4 The percentage of all interactions that were of a sexual nature increased to a peak in late December ($\chi^2 = 61.08$; df = 5, $P < 0.001$).

wrapping his ambulatory legs around it. Tapping movements of his major cheliped on the female's chelipeds accompanied or alternated with jerky movements of the minor cheliped as he attempted to pull the female towards him by her appendages. The female then moved her chelipeds and ambulatory legs and stroked the male's chelipeds. In many cases the male released his hold on the female's ambulatory legs by his minor cheliped but then

Figure 3.5 Typical posture prior to mating, male holds both chelipeds above female while holding her shell in position with his ambulatory legs.

Figure 3.6 Male (92) easing female out of shell in ventral–ventral contact for mating.

continued to pull her towards him with the major cheliped. Both crabs eased partially out of their shells and their ambulatory legs entwined (Fig. 3.6). In this ventral to ventral position the male transferred spermatophores to the female by use of the reduced 5th pereiopods.

In the case of unsuccessful matings the male was observed to drop the female whilst trying to manipulate her shell, or was interrupted by another crab or simply gave up and walked away. Of the 49 attempts by males in *Gibbula* shells only one was successful whereas of the 64 by males in *Littorina* shells 14 were successful ($\chi^2 = 7.83$, $P < 0.01$). In only two cases of completed matings was the female soft from a recent moult and none of the females had eggs on the pleopods at the time of mating. The spermatophores were on the coxopodites of the 3rd and 4th pereiopods of the females i.e. adjacent to the genital openings on the 3rd pereiopods. All matings were between crabs in the same species of shell and there was size assortative pairing (Fig. 3.7).

Ovigerous females are found on the shore from December to May (Elwood and Stewart 1987; Fig. 3.8). This corresponds with the above findings on the time of mating and agrees with other authors who have found a breeding peak for *P. bernhardus* in the winter and spring (Pike and Williamson 1959; Samuelson 1970; Lancaster 1990). Workers studying *P. bernhardus*

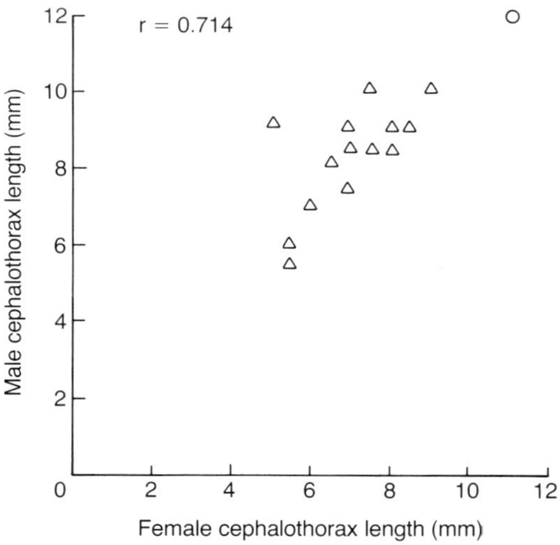

Figure 3.7 Correlation between male and female cephalothorax lengths in mating pairs. ○ = *G. cineraria* pair; △ = *L. obtusata* pair.

sublittorally, however, have found a longer breeding season. Jackson (1913) working at the Isle of Man found females carrying eggs 'almost any time of year', whereas Pike and Williamson (1959) working in the Clyde Sea area and the Isle of Man found ovigerous females from December to September. Samuelson (1970) working in Norwegian waters found ovigerous females from January to July. This indicates that the breeding period may vary with locality, and that sublittoral specimens may have a more extended breeding

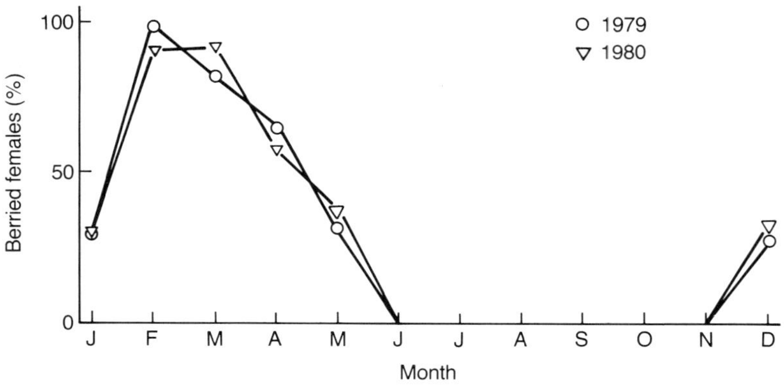

Figure 3.8 Percentage of females that were berried (i.e. bearing eggs) is shown for each month for two years. (From Elwood and Stewart 1987.)

season than littoral animals. Sublittoral specimens typically have more than one clutch of eggs per year and mating may occur at about the time of release of the zoea larvae of the previous batch (Hazlett 1968e). Many littoral crabs also appear to have more than one clutch as indicated by the presence of gravid ovaries whilst eggs are borne on the pleopods; however, it is likely that crabs only have a single clutch in their first year (Elwood and Stewart 1987; Lancaster 1990). Clutch size increases with body size (Samuelson 1970; Fig. 3.9) and large, sublittoral specimens may carry up to 48 000 eggs (Pike and Williamson 1959).

A recent study compared the numbers of eggs produced by females in *L. obtusata* and *Gibbula* shells (Elwood and Marks unpublished). Females in *L. obtusata* shells had more eggs, internally, prior to laying ($P < 0.001$: Fig. 3.10) and, on their pleopods, after laying ($P < 0.001$: Fig. 3.10). There was no apparent loss of eggs during laying because numbers held internally were not significantly greater than those subsequently held on the pleopods (although consistently higher in 8 out of 8 groups in Fig. 3.10).

Data from monthly samples over a two-year period indicate that very small crabs, assumed to be recently metamorphosed from the glaucothoe stage, are present on the shore at most times of the year (Elwood and

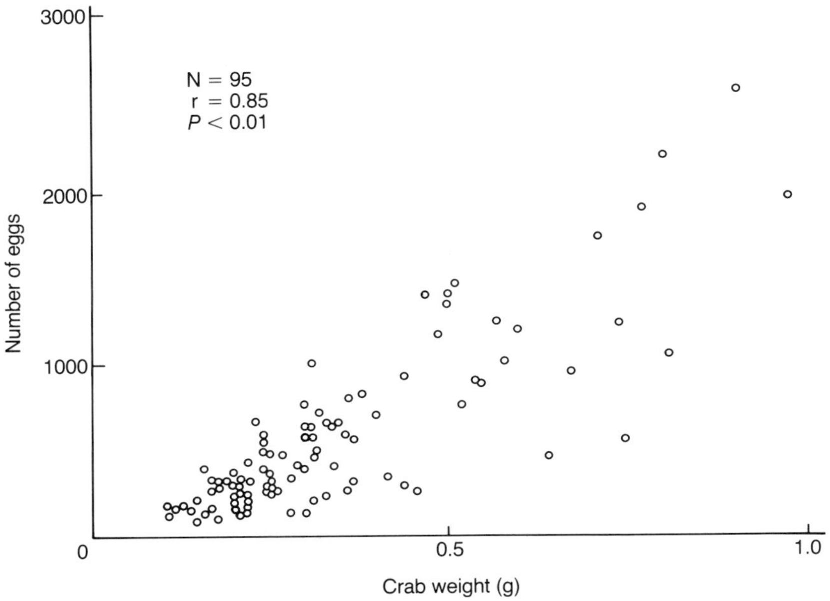

Figure 3.9 Numbers of eggs held on the pleopods of crabs of different weights (weight without eggs). (From Elwood and Stewart 1987.)

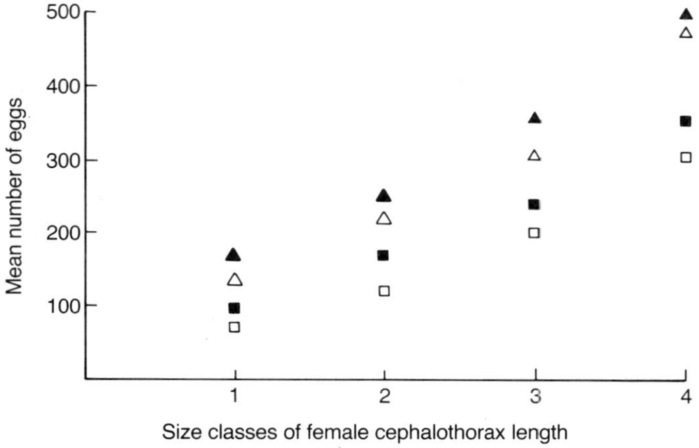

Figure 3.10 Both female size (different size classes) and shell type influence the number of eggs prior to laying (closed symbols) and after laying (open symbols). Crabs in *L. obtusata* shells (triangles), in *G. cineraria* shells (squares). Size class 1 = < 7 mm; 2 = 7–9 mm; 3 = 9–10 mm; 4 = 710 mm.

Stewart 1987; Fig. 3.11). It has been estimated that the female 'incubates' the eggs for approximately 3 months (Pike and Williamson 1959) and that subsequent larval development is approximately 52–62 days (Bookhout 1964). Thus there is a peak in occurrence of small crabs on the shore (Fig. 3.11) some 5 months after the peak in ovigerous females. The presence of very young crabs at other times, however, suggests that they are the product of reproductive efforts by sublittoral animals, which, as stated, produce eggs over a longer period (Elwood and Stewart 1987). It should be noted, however, that survival of glaucothoe is dependent to a large extent

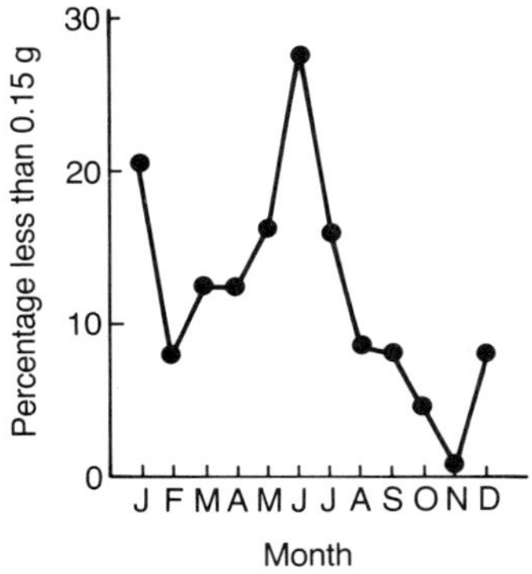

Figure 3.11 The percentage of each sample that weighed less than 0.15 g is shown for two years combined. (From Elwood and Stewart 1987.)

on the availability of small shells. Thus, the variation in numbers of young animals may reflect shell availability as well as the number of metamorphosing larvae.

PREFERENCES FOR PARTICULAR SIZES OF SHELLS

Much of our experimental work on assessments and decisions requires that crabs are offered shells that deviate from preferred size by a known amount. In this way it is hoped to control one aspect of the motivational state of the crab as regards shell acquisition. It is thus imperative to know the preferred size of shell for any crab. The first problem is to quantify shell size in some way. Previous workers have used aperture width, height of shell, internal volume, weight or a combination of these factors (Conover 1978). Any of these are of use in quantifying within a species of shell but these measures may not be suitable for comparing between species. We have measured weights and volumes of each of the main species employed in this study and in each species found high correlations between the two measures (broken or worn shells were not analysed). In our experiments shell weight was used as the measure of size but all shells were checked to ensure that the volume did not deviate widely from the expected value for that weight.

It is apparent from Table 3.1 and the above discussion that the vast majority of the crabs on our study site occupy the two small species of shell and it is these crabs that were used in most of our experiments. In most cases the crabs tested ranged from about 0.15 to 0.45 g. To determine shell size preferences of crabs in this range crabs were offered a range of shells of a single species and the weight of each shell occupied at each observation (spread over several days) was recorded. The average weight of the shells occupied by each crab was taken as the 'preferred' weight of that shell species for that crab. Regression lines were then calculated and used to determine the preferred size of shell over a range of crab sizes (Figs 3.12 and 3.13). Most of our experiments have been conducted only on male crabs to eliminate variation due to eggs being carried on the abdomen by females and the data presented here for crabs offered *L. obtusata* or *Gibbula* shells are for males only. Somewhat surprisingly, however, the possession of eggs does not change the preferred weight of shell for small crabs (Dowds 1983) but some sex differences in shell preference are apparent in larger crabs (see below). An alternative method for assessing shell preferences was employed with larger crabs. In this case crabs, collected in either *N. lapillus* or *L. littorea* shells, were placed for two days in large tanks containing many empty shells of both species. This gave crabs sufficient time to investigate many shells and the shell occupied at the end of this period was taken to be of the preferred species and size. In these cases females tended to occupy heavier, and hence larger, shells of each species.

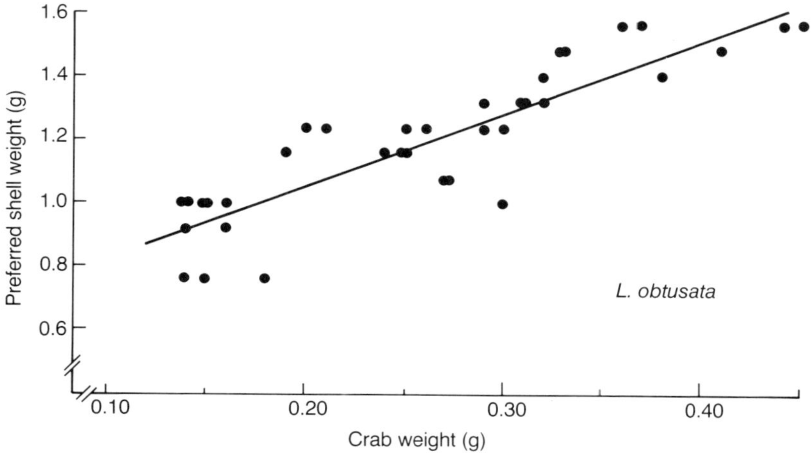

Figure 3.12 Crab weight is plotted against preferred shell weight for *L. obtusata* shells ($y = 0.60 + 2.27\,x$; $r = 0.88$, $n = 36$, $P < 0.01$). (From Jackson 1988.)

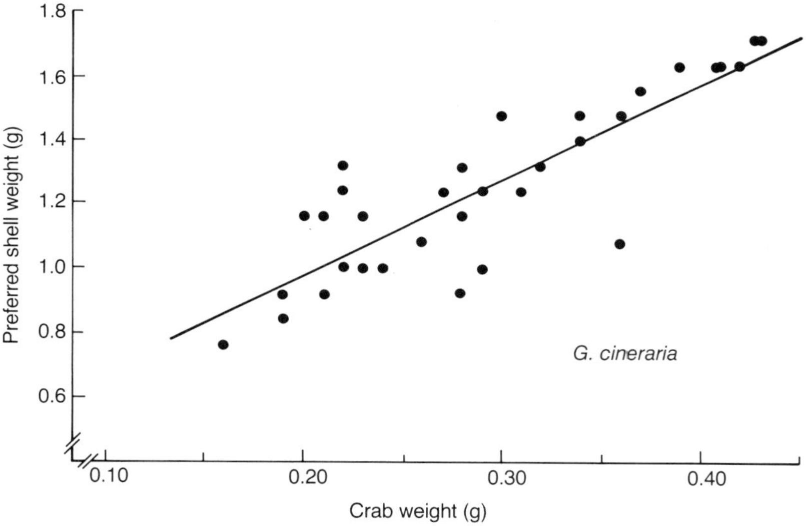

Figure 3.13 Crab weight is plotted against preferred shell weight for *G. cineraria* shells ($y = 0.38 + 2.99\,x$; $r = 0.86$, $n = 33$, $P < 0.01$). (From Jackson 1988.)

PREFERENCES FOR PARTICULAR SPECIES OF SHELLS

The data in Table 3.1 show that a significantly smaller proportion of *Gibbula* shells on the lower shore are occupied than are *L. obtusata* shells ($\chi^2 = 678.1$; $P < 0.001$), suggesting that the latter is the preferred species. This requires direct testing, however, as the lack of empty *L. obtusata* shells might be because they are more subject to movement by wave action and thus swept to the high water mark where they are unavailable to crabs and to our survey. As some aspects of our later work have utilized species preferences as an experimental variable this apparent preference has been tested several times. For example, Elwood *et al.* (1979) collected two groups each of 30 crabs in *Gibbula* shells and two groups each of 30 crabs in *L. obtusata* shells. Each group was placed with 45 shells of the alternative species and the number of crabs in each species was monitored several times a day for 4 days. The numbers in each species fluctuated over the 4 days and the median number in each was calculated. This showed that, on average, 38 out of 60 crabs originally in *Gibbula* shells changed to *L. obtusata* whereas only eight crabs originally in *L. obtusata* moved to *Gibbula*. Clearly the shift to *L. obtusata* shells was greater than the shift to *Gibbula* shells ($\chi^2 = 29.6$; $P < 0.01$) which supports the hypothesis that the former is the preferred species.

It is possible, however, that crabs in *Gibbula* shells may have less adequate shells as regards size irrespective of species preferences. Thus those crabs may equally have preferred to move to a better fitting *Gibbula* shell as to move to one of the alternative species. In an attempt to avoid this problem Jackson (1988) used crabs collected in either *Gibbula* shells ($n = 37$) or *L. obtusata* shells ($n = 39$) and cracked the shells to remove the crabs. Each naked crab was placed with five shells of each species, one of each being the preferred weight for that crab weight (Figs 3.12 and 3.13), the others being 0.08 g or 0.16 g more or less than the preferred. The crabs were again monitored several times daily for 4 days. The initial choice of shell was random with respect to species (Fig. 3.14), but thereafter there was a marked tendency for the crabs to move into the *L. obtusata* shells. For those originally collected in *L. obtusata*, all spent more than half of the time in that species, i.e. preferred them, and only a small proportion of those originally collected in *Gibbula* shells seemed to prefer *Gibbula* shells. These data demonstrate a clear preference by these small crabs for *L. obtusata* shells, which happens also to be the most common species on the shore. It has been noted above that males in *L. obtusata* shells obtained more matings than did those in *G. cineraria* shells in an experimental situation. Furthermore, females in the preferred *L. obtusata* shells had more eggs than did those in the less favoured *G. cineraria* shells (Elwood and Marks unpublished). Thus

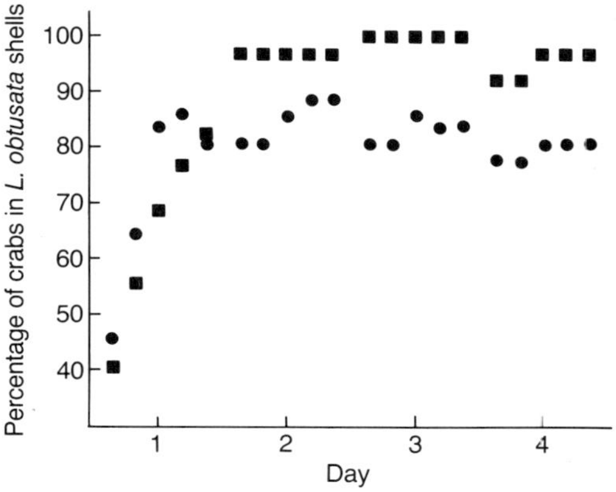

Figure 3.14 Indication of shell species preference shown by the percentage of crabs that were occupying *L. obtusata* shells at each observation time. Crabs originally from *L. obtusata* shells (▪) and *Gibbula* shells (•). (From Jackson 1988.)

the preferences shown by these animals seem to be determined by fitness consequences.

THE EFFECTS OF SEX DIFFERENCES AND EXPERIENCE ON SHELL PREFERENCE

The experiments cited above to determine preference between *L. obtusata* and *Gibbula* shells, apart from revealing a preference for the former, also revealed that the degree of preference was influenced by the shell in which the crab was collected. For example, 22 out of 60 crabs failed to change from their *Gibbula* shells to *L. obtusata* shells whereas only eight out of 60 in *L. obtusata* moved to *Gibbula* shells ($\chi^2 = 7.5$; $P < 0.01$) (Elwood *et al.* 1979). Furthermore, when naked crabs were offered a choice of the two species (Jackson 1988) seven out of 37 originally from *Gibbula* shells spent most time in that species whereas none of the 39 from *L. obtusata* did so ($\chi^2 = 6.0$; $P < 0.02$) (Fig. 3.14). It is thus possible that experience of inhabiting *Gibbula* shells, forced by a shortage of the preferred species, has resulted in a shift in shell preference in some crabs.

This possibility of a shift in preference was also investigated in larger crabs collected in either *N. lapillus* or *L. littorea* shells (Elwood *et al.* 1979). Crabs found in these species were removed from their original shells and offered a choice of two new shells. One of these was of the same species and the same weight as their original and the other was of the alternative species and also of the preferred weight. In this manner 54 crabs from *N. lapillus* and 51 from *L. littorea* were tested. Thirty-two of the crabs from *N. lapillus* shells chose those again whereas 34 from *L. littorea* shells chose those again. Thus crabs offered shells of equivalent weights preferred their original species ($\chi^2 = 6.51$; $P < 0.02$). This preference for the species of shell in which the crab was collected was maintained even when the weights of the shells on offer were varied considerably from the optimum (Elwood *et al.* 1979).

Three hypotheses have been proposed to explain this apparent development of individual preferences. Firstly, a learning process may occur with crabs successfully living in one species learning the features of that species (Elwood *et al.* 1979). Secondly, the shape of the crab may become moulded when the exoskeleton is soft after ecdysis (McLaughlin and Bailey-Brock 1975; Elwood *et al.* 1979). Thirdly, it has been suggested that as *L. littorea* shells grow to a larger size than *N. lapillus* shells the data above may simply reflect larger crabs preferring the largest shells (Blackstone 1984).

The 'moulding hypothesis' has been tested recently using *P. longicarpus* (Blackstone and Joslyn 1984; Blackstone 1985) and *P. bernhardus* as subjects (Elwood and Kennedy 1988). Subjects of both species were maintained in particular species of shell and subsequently their preferences were tested (Blackstone 1985; Elwood and Kennedy 1988) and various morphological features of the animals measured (Blackstone and Joslyn 1984; Elwood and Kennedy 1988). These studies showed that long-term occupation of one particular species did not induce a preference for that species and evidence for morphological change was limited. What is apparent from these studies is, however, that the sexes show different preferences. Within particular size classes *P. bernhardus* males show a preference for *L. littorea* shells whereas females show a preference for *N. lapillus* shells. This occurs for animals collected in those shells on the shore as well as for animals maintained in one species or the other in the laboratory (Elwood and Kennedy 1988). Thus, the original suggestion of the development of individual preferences (Elwood *et al.* 1979) appears to be erroneous; rather there is a sexually determined preference.

ARE CRABS ON THE SHORE IN ADEQUATE SHELLS?

The majority (94%) of the *L. obtusata* shells, which are the preferred species for small crabs, are occupied (Table 3.1) but this still leaves 6% unoccupied. Not all of these unoccupied shells may be available to crabs in less preferred species, however, as some may have only been dislodged from under rocks during collecting. Even if located by a crab these shells may not be of a suitable size for that crab. It appears likely that some crabs are unable to find a sufficient supply of shells to ensure that they always have one of the preferred species and size. Thus we find some in the less desirable *Gibbula* shells and others, although in the preferred species, are either much too large for their shell to enable withdrawal or are too small for their shell and have to carry excessive weight. Crabs may also be found in shells that are damaged to such an extent that they offer little protection. When offered alternative shells in the laboratory or field these crabs readily accept new shells.

This indicates that some crabs on the shore inhabit unsuitable shells. Even so, the crabs on our study site appear to have a better choice of shells compared to those in many other sites in which shells are a severely limiting resource (Chapter 2). The process by which hermit crabs make their choice of shells is examined in the following two chapters. Chapter 4 examines the behaviour of crabs when they encounter shells, i.e. the assessment process, and Chapter 5 examines motivational changes associated with the assessment procedure, i.e. the decision-making mechanism.

CHAPTER 4

Assessment of unoccupied shells

It is clear from the previous two chapters that shells are an important resource and that different shells have different effects on the fitness of the owner. The importance of having an optimal shell may be seen when a crab encounters another shell. The crab typically engages in a series of activities at the end of which it either rejects that shell or moves out of its old shell and into the new one. That series of activities (described more fully below) appears to be a process by which information is gathered that enables the decision of 'accept' or 'reject' to be made, i.e. the crab assesses the quality of the shell.

Shells may vary in many ways so shell assessment is likely to be complex, although it is by no means the most complex example of assessment. Perhaps that title must go to the honey bee swarm selecting a new nest site (reviewed by Seeley 1985). A number of scouts leave the original colony and seek cavities large enough for a honey bee nest. When a cavity is located the scout inspects it inside and out for up to an hour. Apart from the longer time taken, this aspect is not dissimilar to shell assessment by hermit crabs. The honey bee then returns to the hive and advertises the location of the potential new home. The added complexity occurs because many other scouts advertise other potential sites. The vigour of the dance indicates the quality of the potential site and some scouts switch 'allegiance' quickly and others switch after visiting an alternative site. By this means, over a period of several days, the number of sites being advertised is reduced until a 'concensus' is reached. Clearly this is a long, complex process and, although many important features that make a good potential site have been experimentally determined, the final decision-making process is difficult to investigate.

Shell selection by hermit crabs is obviously simpler than selection of nest sites by honey bees, but it still involves a number of activities and a number of decision points. Shells have been shown to be located from a distance by vision (Reese 1963; Kinosita and Okajima 1968). Experiments with *P. samuelis*, for example, in which either black or white shells were offered against either black or white backgrounds indicated that black shells against white backgrounds were the most effective stimuli to elicit approach and that crabs with their eyes ablated showed no such preference (Reese 1963). The amount of information gathered visually, however, appears to be limited as crabs did not discriminate between the shapes of preferred or unpreferred species by vision alone (Reese 1963); discrimination required contact with the shell. The land hermit crab, *Coenobita rugosus*, on the other hand, shows a clear ability to discriminate between different sizes of shells by vision (Kinosita and Okajima 1968) and *Pagurus bernhardus* appears to have similar abilities. *P. bernhardus* also appears to be able to gain information about different shell species prior to contact with the shells (Chapter 5).

P. bernhardus initiates shell assessment by approach followed by contact with the shell with the antennae or, less commonly, with a walking leg or cheliped (Fig. 4.1). The shell is grasped and the exterior explored by moving the chelipeds over the surface (Fig. 4.2), the shell being turned during this process (Fig. 4.3) so that the aperture is uppermost. The interior of the shell is then explored by inserting chelipeds and walking legs (Fig. 4.4). This aperture exploration may alternate with exterior examination, but the crab may occasionally stop active investigation and hold the new shell without moving. The shell may be rejected at any stage or, after investigation, the crab may grasp the lip of the aperture with its chelipeds, release its abdominal grip on the old shell and swing the abdomen swiftly into the new shell

Figure 4.1 Approach by crab in a *G. cineraria* shell to an *L. obtusata* shell.

Figure 4.2 External investigation. The crab in a *G. cineraria* shell holds on to the *L. obtusata* shell while moving its chelipeds over the surface.

Figure 4.3 Turning the shell. The crab turns the shell over to investigate the interior.

Figure 4.4 Internal investigation. The crab investigates the aperture and interior of the shell.

(Fig. 4.5). The crab then turns to grasp the shell just vacated which it then investigates in a manner similar to that just described (Fig. 4.6). In addition, the crab performs a number of activities concerned with the new shell. The crab may withdraw into the new shell, often repeatedly in a vigorous manner which Reese (1963) calls 'snap' withdrawal (Fig. 4.7). It may swing the new shell off the substrate and hold it aloft for a second or two (Fig. 4.8), or it may run the walking legs over the surface of the new shell (Fig. 4.9). Sometimes these activities are also seen with the original shell during the early stages of shell investigation prior to the crab moving into the new shell. After moving into the new shell and investigating the old the crab may either move back to the original or move off in the new shell. The onset of each of these activities may be viewed as a discrete decision point (Fig. 4.10), each decision being whether or not to continue the process of investigation.

Before we analyse this sequence in more detail we will make some assumptions that simplify the analysis. Firstly, we may assume that all activities incur a cost to the performer in terms of time, energy or risk and that this cost should be offset by some benefit from the performance of the activity. Natural selection should optimize the benefit-to-cost ratio of sequences of behaviour (McCleery 1978) and thus the cost of investigating a shell prior to moving in must be offset against some benefit, otherwise the crab should

Figure 4.5 Changing shells. The crab rapidly moves from one shell (right) to the other (left).

Figure 4.6 The crab re-investigates its old shell.

Figure 4.7 Withdrawn. The crab withdraws fully into the shell. The crab in the photograph is not quite fully withdrawn.

Figure 4.8 Shell twirling. The crab lifts the entire shell off the substrate in a twirling motion.

Figure 4.9 Feeling the new shell. The crab reaches round the back of the shell it is occupying with its legs.

simply move into the shell in order to assess its quality. What then is the benefit of engaging in shell investigation prior to moving? Presumably moving into an unknown shell involves some risks; for example (i) the shell may be occupied and a fight may ensue, (ii) another crab may grasp the evacuated shell, thus leaving the crab which just left it with a shell of unknown quality, or (iii) the risk of predation may be increased during the change or the change back if the new shell was unsuitable. We therefore assume that the benefit is in the form of information concerning the availability and suitability of the shell. This information, however, is not perfect, as evidenced by the number of crabs that return to the original shell after moving to the new shell. We may expect, however, that the number of such double changes (mistakes?) is reduced by the information gained during investigation. Shell investigation thus appears to be a mechanism for information gathering. It is also an assessment and decision-making process that is particularly amenable to experimental manipulation (e.g. Reese 1963; Kinosita and Okajima 1968; Elwood and Stewart 1985; Neil and Elwood 1985).

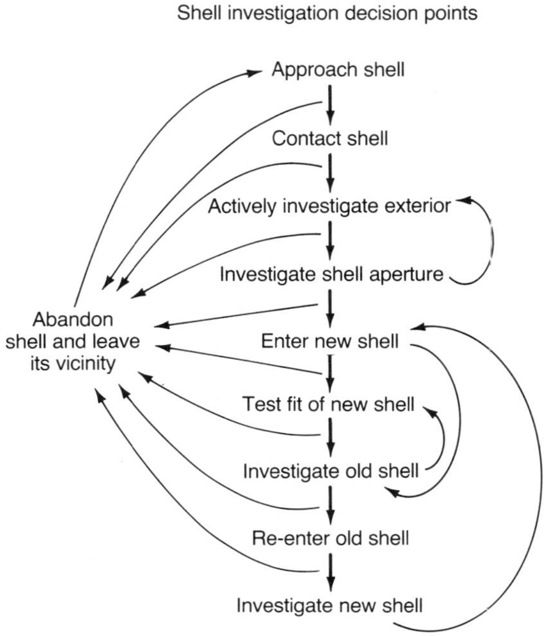

Figure 4.10 The main activities during an investigation are indicated in the order in which they typically appear (from top to bottom). The thin arrows indicate other common transitions between activities.

IS INFORMATION GAINED DURING SHELL INVESTIGATION?

Whether or not information is gained during shell investigation may be ascertained by varying the potential gain from acquiring the new shell and determining if that influences the likelihood of shell change. If crabs are more likely to change following investigation when a large gain may be made, then they must have gathered that information without moving into the new shell; that is, they must have made an assessment. This possibility has been examined in our laboratory (Elwood and Stewart 1985). We have used standard methods in all our studies and this section introduces our basic methodology.

Only undamaged and unparasitized male crabs collected in *L. obtusata* shells were used, thus reducing variation due to sex differences and shell preferences (Neil and Elwood 1985). In the laboratory, crabs were housed in aerated sea water at approximately 6–8 °C with hours of illumination appropriate for the time of year. The day prior to observation crabs were cracked out of their shells in a bench vice (this does not damage the crabs!), gently dried with a paper towel and weighed. Each crab was then placed in a separate 11 cm diameter crystallizing dish filled with 2–3 cm of sea water.

In this experiment (Elwood and Stewart 1985) 90 crabs remained naked (0% group), 90 were given a *L. obtusata* shell that was 50% of the preferred weight (as determined in a preliminary experiment) (Fig. 3.12; 50% group), 90 were given a shell of the preferred weight (100% group) and 90 were given a shell that was 150% of the preferred weight (150% group). All crabs that were given shells entered them within 2 hours, and most did so within a few minutes. The following day (16–18 hours later) the crystallizing dish was placed behind a one-way Perspex partition and an empty shell (filled with sea water to expel air) was placed in the middle of the dish with the aperture downwards. This new shell was either 50%, 100% or 150% of the preferred weight for that crab. In this manner 12 experimental groups (in a 3 × 4 design) were established, each containing 30 crabs. The activities of each crab were recorded over a 30-minute period or, if the crab was investigating the new shell, until the crab either entered it or rejected it. The data were recorded directly onto computer.

Of the 360 crabs observed, 292 (81%) approached and made contact with the offered shell, 258 (72%) initiated an active investigation of the shell and 211 (59%) entered the shell. Major decisions that crabs appear to be making prior to shell entry are thus whether to (i) approach and make contact, (ii) actively investigate and (iii) enter the offered shell. In the following section the factors that influence these three decisions are examined, particularly the two major variables of this experiment: (i) the size of shell in possession and (ii) the size of shell on offer.

Whether to approach and make contact

This was significantly influenced by the shell inhabited, with naked crabs being the most likely to approach and make contact. Fewer crabs per group made contact as the size of the shell in possession increased (Fig. 4.11a). If approach and contact with another shell is a reflection of a crab's motivation to seek and change shells this is not the expected pattern. It would be predicted that crabs in their optimum shell size (100% group) would have the lowest motivation to change shells and thus the lowest motivation to approach other shells. That crabs in the 150% group approached the least (although not significantly less than the 100% group) suggests that perhaps the crabs in these excessively large shells had a slightly reduced mobility due either to poor fit or excessive shell weight.

The quality of the shell on offer also influenced whether or not crabs approached, with crabs being most likely to approach optimal shells and least likely to approach the 50% shells (Fig. 4.11b). This indicates that the crabs must initiate their evaluation of shell quality prior to contact with the

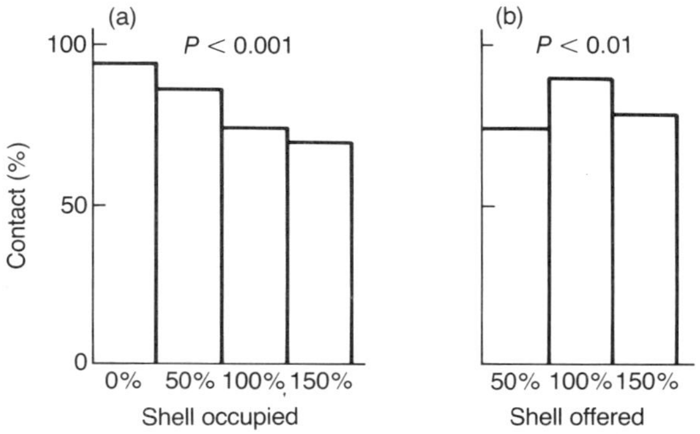

Figure 4.11 The percentage of crabs making contact with the offered shell depending on (a) the size of the occupied shell and (b) the size of the offered shell. Shell sizes are expressed relative to the size preferred by the crab. (From Elwood and Stewart 1985.)

shell; presumably this initial, distant evaluation is visual as has been shown in *P. samuelis* (Reese 1963) and *C. rugosus* (Kinosita and Okajima 1968).

Whether to investigate

Crabs in optimal shells were the least likely to actively investigate the shell (external, aperture, or both) once contact had been made (Fig. 4.12a) and this is in agreement with the prediction that these crabs should have the lowest motivation to change shells. There was, however, no significant difference in the probability of active investigation dependent upon the quality of the shell on offer (Fig. 4.12b), which indicates that little, if any, further assessment of shell quality takes place between the initial visual assessment and contact and the onset of active investigation.

A study by Neil and Elwood (1986) found that there was a significant relationship between the size of the occupied shell and whether or not the animals which began external investigation proceeded to aperture investigation. Crabs in good quality shells were the most likely to stop investigation prior to aperture investigation, and the quality of the shell on offer did not appear to influence this decision. Thus external investigation, somewhat surprisingly, does not appear to enable a crab to gather much information about shell quality or at least the information is not utilized in the decision to start aperture investigation. That the latter explanation is the more likely, however, may be deduced from experiments described in the following chapter.

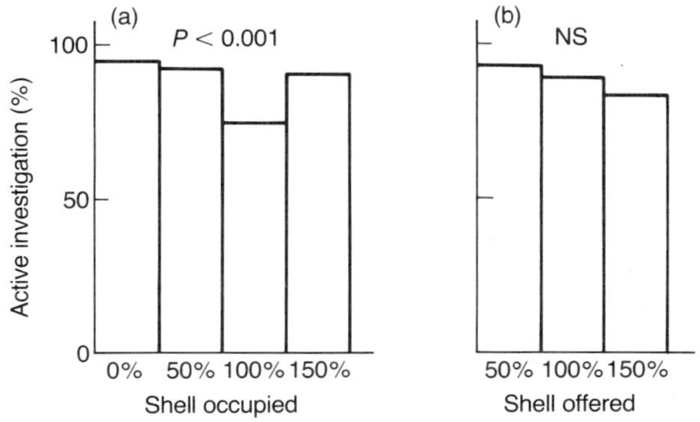

Figure 4.12 The percentage of crabs that made contact continuing to active investigation depending on (a) the size of the occupied shell and (b) the size of the offered shell. Shell sizes as for Fig 4.11. (From Elwood and Stewart 1985.)

Whether to enter the new shell

Of the crabs that began active investigation, those in optimal shells were the least likely to enter the offered shell whereas all naked crabs that investigated the offered shell entered it (Fig. 4.13a). Furthermore, of those crabs that began active investigation those offered optimal shells were the most likely to enter those shells whereas those offered 50% shells were the least likely (Fig. 4.13b).

It is clear from these data that hermit crabs utilize information concerning the shell they have in possession in the major decisions of whether or not to

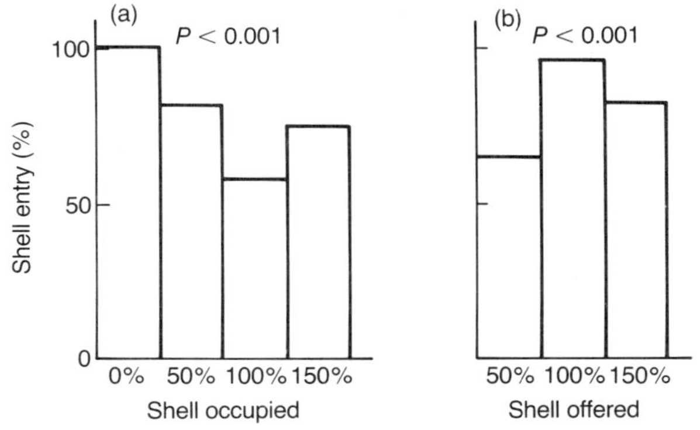

Figure 4.13 The percentage of crabs that actively investigated continuing to shell entry depending on (a) the size of the occupied shell and (b) the size of the shell on offer. Shell sizes as for Fig 4.11. (From Elwood and Stewart 1985.)

approach and contact, to investigate the exterior and interior and to enter the offered shell. Information concerning the quality of the offered shell, however, apparently is only utilized in the decisions of whether or not to approach and contact and, primarily, whether or not to enter the shell.

The data shown in Figs 4.11–4.13 are for the percentages of those crabs that reached one stage and continued to the next stage of investigation. The cumulative effect of these decisions is shown in Fig. 4.14 which indicates the number of crabs in each of the twelve groups that entered the offered shells. This again shows that crabs in optimal shells have a low likelihood of entering new shells whereas those offered optimal shells are the most likely to enter the new shells. Crabs thus appear to estimate the gain that may be made from obtaining the new shell and this apparently entails a comparison of the relative qualities of the shell on offer and that in possession. However, the information gained might also be used in ways that do not involve comparison (see Chapter 5). It is thus clear that shell assessment by hermit crabs is a means of gathering information.

USE OF SPECIFIC APPENDAGES DURING INVESTIGATION

During shell investigation crabs may engage in any of a number of acts. We have classified seven distinct acts during aperture investigation depending upon which of the various appendages are inserted into the aperture (Table 4.1). In addition, external exploration and passively holding the shell also occur, making a total of nine acts recorded during shell investigation (we include turning the shell as part of external exploration). These acts are not all used with the same frequency and some transitions between acts are more common than others. Furthermore, crabs differ in their use of specific acts and in the total number of acts depending upon the quality of both the occupied shell and that being investigated. Crabs investigating large shells (150%), for example, are more likely to use the major cheliped either alone

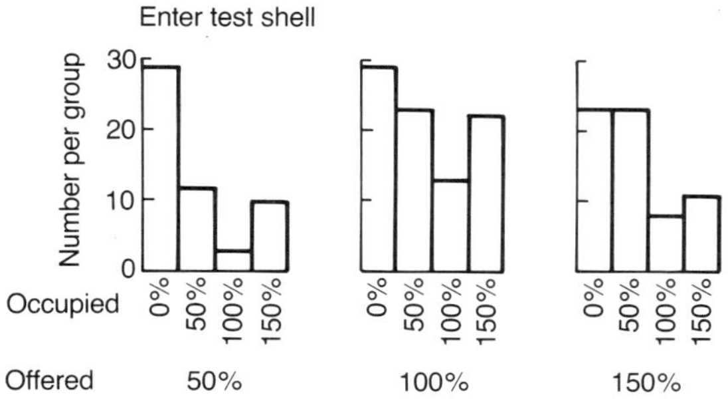

Figure 4.14 The number of crabs in each group that entered the offered shells. Shell sizes as for Fig 4.11. (From Elwood and Stewart 1985.)

Table 4.1. Activities recorded during shell investigation

Behaviour	*Description*
External exploration	Crab holding shell with walking legs, actively moving chelipeds over shell surface
Hold	Crab grasps shell but no active investigation
Aperture exploration:	
Leg	One or more walking legs inserted into aperture
Minor + leg	One or more walking legs plus minor cheliped inserted into aperture
Major + leg	One or more walking legs plus major cheliped inserted into aperture
Both chelipeds + leg	One or more walking legs plus both chelipeds inserted into aperture
Major	Major cheliped inserted into aperture
Minor	Minor cheliped inserted into aperture
Both	Both chelipeds inserted into aperture

Table 4.2. The percentages of crabs using specific investigatory activities when offered shells of different sizes. *P* values relate to χ^2 tests performed on raw data

	Size of offered shell			
Activity	*50%*	*100%*	*150%*	P *value*
Major cheliped	61%	69%	80%	<0.05
Major cheliped + legs	0%	10%	18%	<0.01
Both chelipeds + legs	15%	43%	43%	<0.01

or in combination with other appendages (Table 4.2). This may be because the major cheliped is better able than the minor cheliped to reach deep into large shells in order to assess some aspect of potential fit.

More surprising is the finding that crabs in different sizes of shells use different appendages in shell investigation (Table 4.3). However, much of the variation is due to the very short investigations exhibited by naked crabs, which results in few crabs using each combination (see below). This problem of having some short sequences may be overcome by examining the use of appendages in the final act prior to shell entry (Elwood and Stewart 1985). Housed crabs almost invariably use the major cheliped in the final act of investigation whereas naked crabs are more likely to use the minor cheliped as the final act (Fig. 4.15). Perhaps naked crabs are at greater risk during shell investigation and keep their major cheliped ready for defence (Elwood and Stewart 1985).

As stated above, housed crabs use significantly more acts in shell assessment than do naked crabs (mean number of acts 8.09 and 2.55, respectively). The main transitions between investigatory acts for housed crabs are shown

Table 4.3. The percentage of crabs, housed in different size shells, using specific investigatory activities when occupying shells of different quality. *P* values refer to χ^2 tests performed on raw data

	Size of occupied shell				
Activity	*0%*	*50%*	*100%*	*150%*	P *value*
Major cheliped	33%	95%	90%	95%	<0.01
Minor cheliped	47%	66%	48%	70%	<0.05
Both chelipeds	18%	95%	100%	95%	<0.01
Minor cheliped + leg	0%	14%	21%	44%	<0.001
Both chelipeds + leg	0%	41%	76%	70%	<0.001

in Fig. 4.16a which indicates that following external exploration the chelipeds are employed alone, or in combination. Also indicated is the large number of transitions between insertion of 'major', 'both' and 'minor' chelipeds. Data for naked crabs (Fig. 4.16b), however, indicate that transitions between these three acts are rare.

The normal sequence of aperture investigation by housed *P. bernhardus* may be deduced from Fig. 4.16a. First, either the major (right) or minor (left) cheliped is inserted, followed by inserting whichever cheliped remains free. A walking leg may then be inserted and then removed, followed by removal of the minor cheliped. As stated above, the major cheliped is usually the final appendage to be withdrawn from the aperture prior to shell entry. In *P. longicarpus* the right cheliped is most commonly the first appendage to be inserted, followed by the left cheliped and then the first right walking leg (Scully 1986). In other species of crab the walking legs are less likely to be

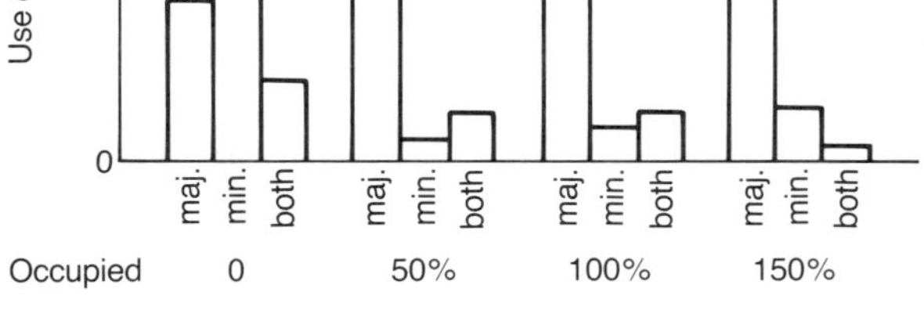

Figure 4.15 The use of chelipeds in the final act prior to shell entry depending on the size of the occupied shell. (From Elwood and Stewart 1985.)

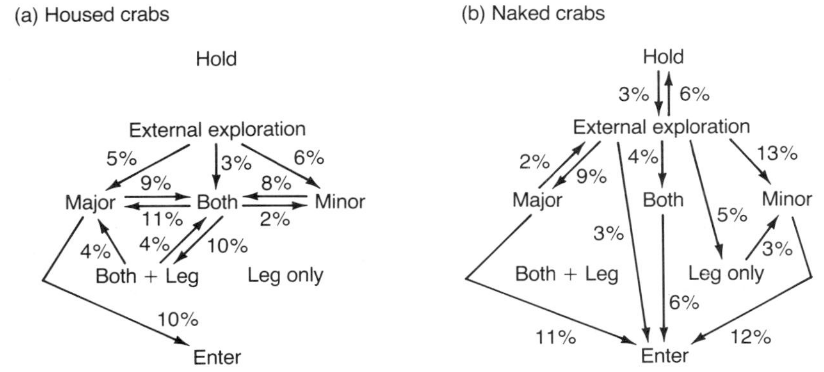

Figure 4.16 The most frequent transitions between acts during shell investigation by crabs in shell (a) and by those that are naked (b). For (a) 127 crabs engaged in 1012 acts (mean 8.09) and for (b) 80 crabs engaged in 204 acts (mean 2.55). Percentages refer to percentages of all transitions.

inserted except when one or more of the chelipeds are missing, apparently in an attempt to compensate for this loss (Reese 1963; Kinosita and Okajima 1968). With uninjured crabs, however, it is not clear whether the walking leg is used in information gathering; it has been suggested that it may simply be easier to insert it into the aperture rather than to hold it outside (Scully 1986).

It seems, however, that insertion of the chelipeds provides information concerning the quality of the shell and how this is achieved has been studied using two main methods. Firstly, by the removal of one or both of the chelipeds (Reese 1962; Kinosita and Okajima 1968; Scully 1986) in which case there was compensatory use of other appendages. Discrimination of preferred shell species was still possible with both chelipeds ablated (Reese 1962) although more detailed study of the land hermit crab, *Coenobita rugosus*, showed that removal of both chelipeds reduced the ability of crabs to measure the diameter of holes (Kinosita and Okajima 1968). The second method involved less drastic modification of one or both chelipeds (Kinosita and Okajima 1968). Covering the chelipeds and walking legs with a thin coat of wax so that movement was not impaired did not influence the measuring ability of the crab, which indicates that surface receptors are not involved. Changing the shape of the chelipeds, however, did have marked effects. If the pincers on the major cheliped were lengthened by addition of wax then the depth of hole required to elicit abdomen insertion increased, whereas if the pincers were shortened, the depth of hole required was decreased. These experiments confirm the role of the chelipeds in assessing the depth of the aperture and other experiments confirm their role in assessing diameter. Wax, added to either the inner or outer edge of the left (major) cheliped, increased the diameter of the hole required to elicit abdomen insertion whereas addition to the right (minor) cheliped had no effect. It thus appears that the major cheliped is normally the primary organ used to assess both depth and diameter in *C. rugosus*.

Observations made on crabs investigating transparent holes, however, indicated that the chelipeds were moved 'in peculiar ways' (Kinosita and Okajima 1968). This observation led to a series of experiments that examined the role of these movements in shell assessment. Wax was applied to the meropodite of the right cheliped, to reduce movement of the carpopodite–meropodite joint, of crabs that had had the left (major) cheliped removed (Fig. 4.17a). Crabs without wax were able to make normal discrimination of

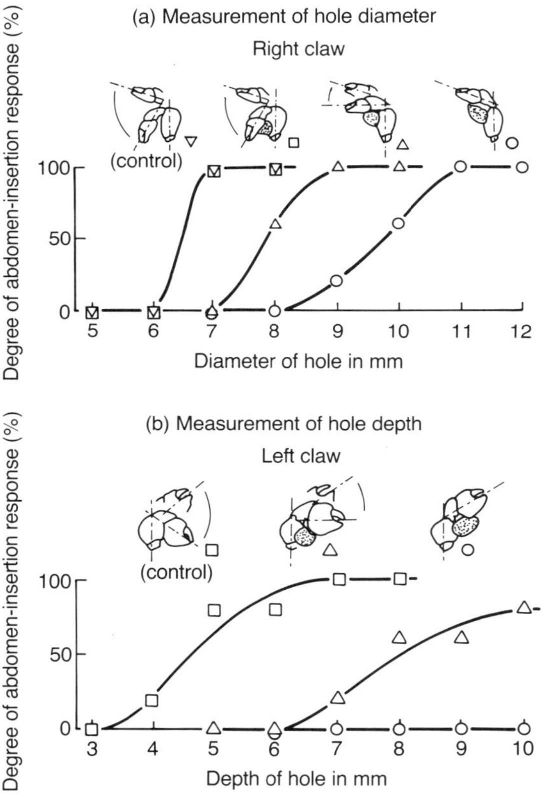

Figure 4.17 Effect of restriction of articular movement of claw on the ability to measure hole diameter (right claw: a) and depth (left claw: b). Yes-or-no choice tests, mean of five successive determinations. Degrees of restriction of movement at carpopodite–meropodite joint are shown in the accompanying diagrams. (From Kinosita and Okajima 1968.)

hole diameter whereas those in which movement was restricted required larger diameters to elicit abdomen insertion (Fig. 4.17b). Similarly in crabs with both chelipeds, restriction of movement of the left one made depth of the hole more difficult to discriminate and increased the depth required for abdomen insertion. Indeed, if the carpopodite–meropodite joint was prevented from closing to less than about 120° then abdomen insertion was not elicited. Thus it seems that assessment of the shell aperture is complex, with the chelipeds used partly as 'static rulers' in measuring depth and width but the size of the concavity is also assessed by the ease with which the chelipeds are moved to-and-fro within the hole. Other experiments eliminated vision as a possible means of assessing the aperture (Reese 1963; Kinosita and Okajima 1968).

DEALING WITH OBSTRUCTIONS IN THE APERTURE

Before considering motivational models of decision making by hermit crabs we should consider how crabs cope with obstructions within the shell aperture. During our experiments (e.g. Elwood and Stewart 1985) and those of numerous other workers (e.g. Reese 1963; Scully 1986) crabs have been observed either examining sound clean shells or shells in which the aperture was blocked with resin or cement. This reflects the experimenters' desire to eliminate unwanted variables caused by broken shells or epibionts. In natural conditions, however, shells will be encountered that may have objects such as sand or gravel within the aperture that the crab would have to remove to enable shell entry. That crabs do remove objects has received brief mention (Reese 1963) but has only recently been the subject of experimental analysis (Elwood and Adams 1990).

In our experiments crabs were housed in *L. obtusata* shells 50% of the preferred weight (Fig. 3.12) and then offered optimal weight shells that (i) were normal empty shells, (ii) had a small quantity of sand in the aperture, (iii) had a dental cement and sand mixture fixed deep within the aperture so that cheliped insertion was still possible, or (iv) had a piece of gravel within the aperture. All shells were offered aperture uppermost.

The probability of approach and approach latencies were similar in the four groups but the actions seen during aperture investigation differed markedly depending upon the obstruction. After inserting one or more chelipeds into the aperture all crabs investigating a shell with sand (17/17) rotated the shell and, in doing so, poured the sand out. Fewer than half the crabs investigating shells with gravel (8/17) or dental cement/sand (7/17) rotated the shell and rotation was rare with crabs investigating empty shells (3/16). Rotation was not the only method of removing sand, however, as crabs scraped out sand by alternate use of the major and minor cheliped and this resulted in considerably more cheliped insertions than in the other

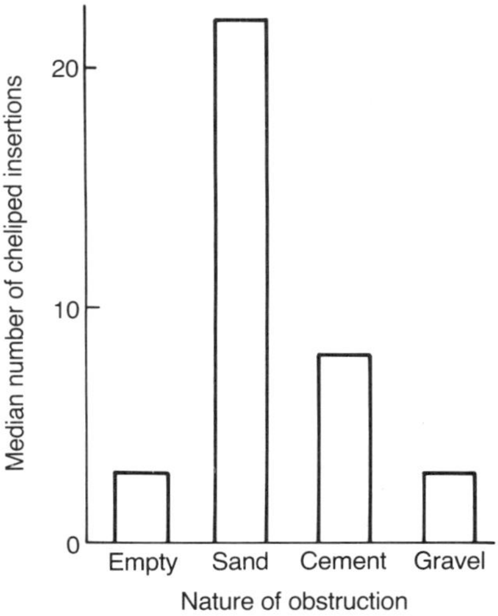

Figure 4.18 The median number of cheliped insertions is shown for the four experimental groups. The 'sand' group showed more insertions than each of the other three groups (Mann-Whitney, $P < 0.002$ for all three comparisons) and the 'cement' group showed more insertions than the 'empty' or 'gravel' groups (Mann-Whitney, $P < 0.02$ for both). (From Elwood and Adams 1990.)

groups (Fig. 4.18). Crabs also inserted chelipeds more often when presented with the cement blockage than with empty shells, but those with gravel in the shell did not differ from those with empty shells in this respect. However, crabs attempted to remove gravel by picking it up in the chelipeds or, failing that, by rotating the shell. As a result of these attempts to remove obstacles crabs took longer to enter shells containing sand or gravel; cement-filled shells were rejected.

This experiment indicates the perceptual abilities of the crabs. They appear to perceive quickly whether or not there is an obstacle, whether or not it is moveable and, if it is moveable, whether it is sand or gravel. Recent experiments on *P. geminus* involved offering sand-filled dextral or synstral shells, and found that crabs could rotate the shells in the appropriate direction to remove the sand (Imafuku, personal communication).

SUMMARY

Hermit crabs approach shells and inspect them visually and then gather further information by manipulating them with their appendages. This involves manipulating the exterior and interior of the shell. If the crab is currently in a poor quality shell and if the new shell is of good quality the crab is more likely to approach the new shell and more likely to accept the new shell. The use of particular manipulatory activities is influenced by the size of the occupied shell and the size of the shell being assessed. The crab may also detect the nature of sand or small stones within the shell aperture and use activities, specific to the nature of the obstruction, to remove those obstructions.

CHAPTER 5

Assessment and motivational change

We have seen in Chapter 4 that when a hermit crab changes to a new shell it normally selects one that is of better quality than the one formerly occupied. In this chapter we examine mechanisms by which this choice is made. Reese (1963) suggested that the motivational process involved in shell selection 'implies a comparison of the total releasing value of two or more stimulus objects either simultaneously or in time'. From his experiments Reese (1963) identified various 'sign stimuli', each of which elicited a 'fixed motor pattern' (Fig. 5.1). The quality of the shell being investigated influenced both the duration and intensity of each motor pattern. Information, resulting from each activity, was viewed as feeding into a 'shell-evaluating centre'. If the crab located a higher quality shell than the one currently occupied then that higher quality was perceived by the crab during investigation and also upon moving into the shell. The information assimilated was viewed as influencing 'appetitive searching behaviour' (Reese 1963). Thus if the crab had moved out of a low-quality shell to one of higher quality it subsequently would have a low tendency to examine the low quality shell that it had abandoned. Selection of a particular shell was viewed as 'a reduction of responsiveness to other shells rather than by any direct comparison between the shells' (Reese 1963). Reese found that a crab would always move into a shell that it had started to investigate. However, observations in our laboratory indicate that investigation of the new shell does not inevitably lead to moving in (Chapter 4). Clearly the decision whether to occupy the shell is based upon information gained during early stages of investigation, and this information appears to influence the duration of particular investigatory activities (Reese 1963; Kinosita and Okajima 1968). The duration of particular activities and the consequent timing of particular decisions are fundamental to our understanding of the motivational processes involved in selection behaviour.

That the shell in possession influences the 'appetitive searching behaviour' is shown not only by the probability of a crab approaching and contacting another shell (Chapter 4) but also by the lapse of time prior to approach once

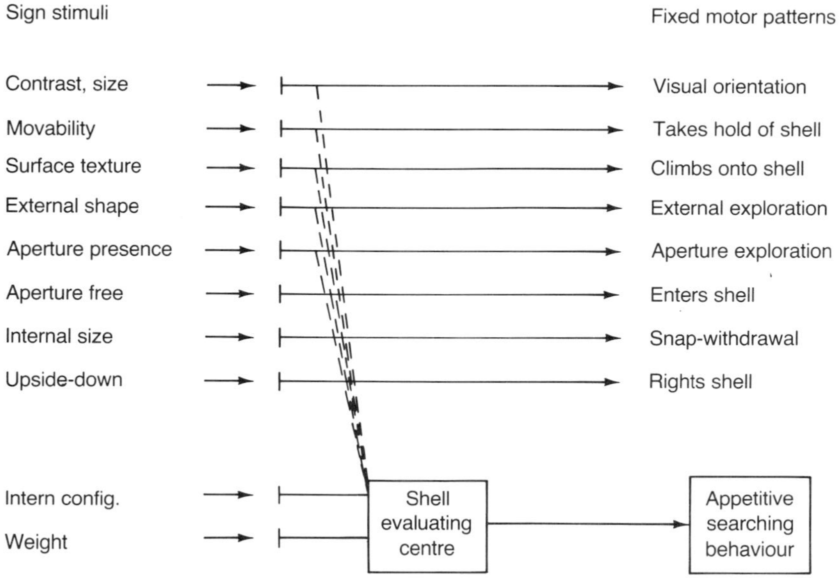

Figure 5.1 A possible scheme to explain the process of hermit crabs selecting shells suggests that information from a variety of sources (sign stimuli) lead into a 'shell evaluating centre' as well as eliciting specific acts. See text for details. (From Reese 1963.)

the animal is introduced to the observation chamber. Neil and Elwood (1986), for example, showed that the poorer the quality of the occupied shell, the sooner did the crab move towards and contact an offered shell. In experiments in which the quality of shell on offer was varied the results have been mixed. In one case (Jackson 1988) the latency to approach was influenced by the visual stimulus of the shell, whereas in another it was not (Neil and Elwood 1986). Furthermore, investigation of the exterior of a preferred species of shell was of shorter duration than was the case with a less preferred species (Jackson and Elwood 1989b).

It is in aperture investigation, however, that we see the most dramatic effects on the duration of activities. In the experiments of Elwood and Stewart (1985), for example, there was a strong negative correlation between the number of crabs in each experimental group that proceeded to aperture investigation and the time spent in that activity (Fig. 5.2). The number of crabs per group commencing an activity provides a measure of the average level of motivation for acquiring a new shell *at the start* of that activity, so the above result indicates that the higher the motivation to change (or, in

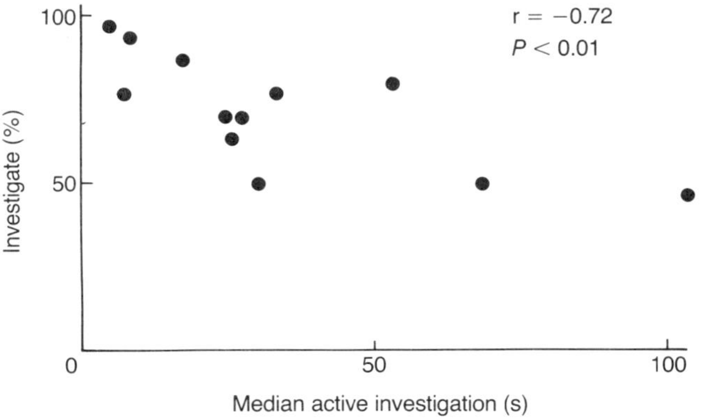

Figure 5.2 A negative correlation was found between the percentage of each of the experimental groups which investigated and the duration of aperture investigation prior to shell entry. (From Elwood and Neil 1986a.)

functional language, the perceived gain at that time), the quicker the shell is accepted.

This result might be explained by the common observation that highly motivated animals tend to move towards a goal more quickly. An alternative explanation, however, is that crabs in some situations might take longer to gather the information necessary for them to make the decision to enter the shell, and we explore this possibility below. We assume two fundamental points: first, that during the shell assessment procedure the information gained alters the motivational state of the animal and second, that the change in motivational state determines when the decision is made concerning whether to enter or reject the shell.

A MOTIVATIONAL MODEL OF SHELL ASSESSMENT

A crab assessing a shell has three options: it may continue investigating the shell ('investigation' consisting of several acts, see Table 4.1), it may reject it or it may move into it. Rejection or moving into the shell are visible decision points; which one is selected appears to be influenced by the relative quality of the two shells but that does not necessarily imply a direct comparison (Reese 1963). As a first step in developing a motivational model of this decision-making process we may view two inputs, one concerning the quality of the shell in possession (poor shell = high input) and the other concerning information gained whilst examining the new shell (good shell = high input) (Fig. 5.3). These will combine in some way to form a single output that will enable the crab to make the decision to move into or reject the

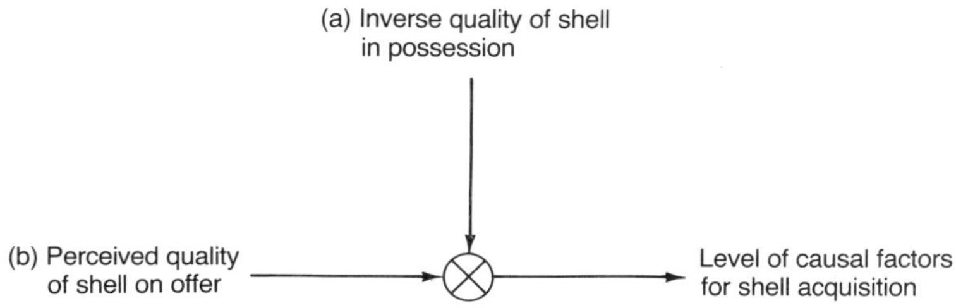

Figure 5.3 The level of causal factors for shell acquisition is the result of two inputs: (a) the quality of the shell in possession (increasing input with decreasing quality), and (b) the quality of the shell being investigated (increasing input with increasing quality).

new shell. Thus, although many aspects of the quality of both shells may influence this decision, we assume stimulus summation to create a single output of causal factors for shell acquisition. This single output may be represented as a single dimension in causal factor space (Elwood and Neil 1986a).

As we are able to reduce the various inputs of information to a single output (Fig. 5.3) of the level of causal factors for shell acquisition, we may plot changes in that level over the course of a shell investigation (Fig. 5.4). Which of the three activities (investigation, rejection or moving in) is selected at any one time will be dependent upon the level of causal factors so that if this level

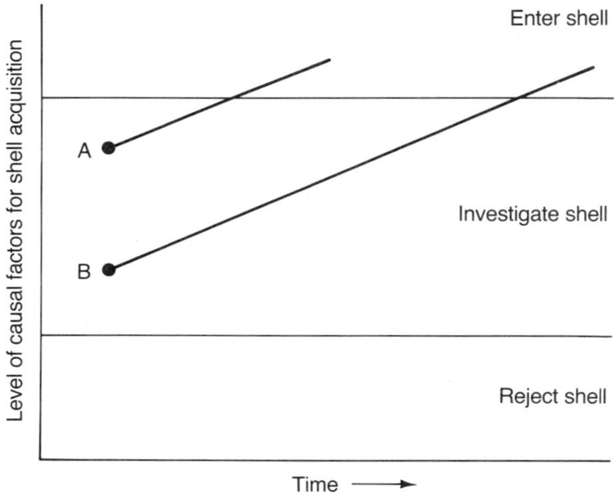

Figure 5.4 Point A represents level of causal factors for crab in a poor shell whereas point B represents level of causal factors for crab in a good shell. Both encounter shells offering the same relative improvements but crab A accepts the shell earlier than does crab B. (From Elwood and Neil 1986a.)

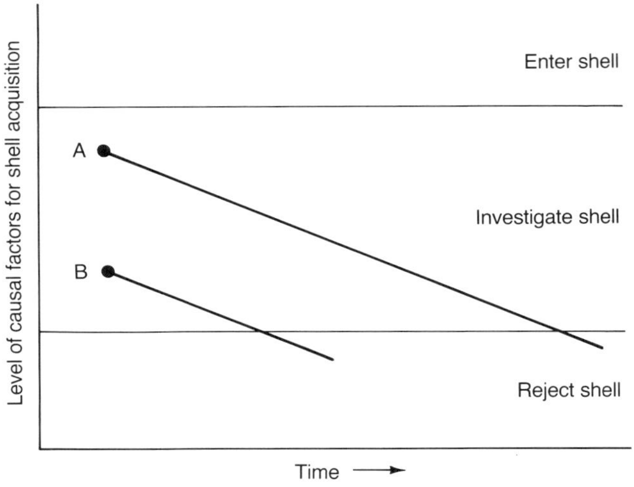

Figure 5.5 Point A – crab in poor shell. Point B – crab in good shell. Both encounter shells which are of same relative loss in shell quality but crab B rejects the shell earlier than crab A. (From Elwood and Neil 1986a.)

rises above a certain threshold the crab will move into the shell or if it falls below a lower threshold it will reject the shell (Fig. 5.5).

First let us consider the initial level of causal factors at the start of shell investigation. This will be determined by the information that the crab has at that time and clearly this will be set primarily by the quality of the shell in possession. A poor shell in possession will determine a high level of causal factors for shell acquisition. The level of causal factors may also be influenced by any distant evaluation of the quality of the shell about to be investigated. We would presume the former to be of more importance at this stage due to the paucity of information concerning the offered shell (Chapter 4). Thus the initial tendency to move to the new shell may be near to the threshold for shell entry for crabs possessing poor shells and near to the threshold for shell rejection for crabs with good shells (respectively points A and B, Fig. 5.4). These points may be viewed as corresponding to animals with high or low motivation to change shells *at that time*.

As time passes the level of causal factors is predicted to change in a direction dependent to a large extent on the information gathered during shell investigation. If the shell being investigated is of good quality the point will move upward toward the shell entry threshold, whereas if it is of poor quality (for example, if it is an unpreferred species, too small, or if the aperture is blocked) the trajectory will be downward toward the rejection threshold. The shell will be either accepted or rejected when one of these thresholds is crossed. If the initial point indicates a high motivation to change shells, i.e. the crab is in a poor shell (point A, Fig. 5.4) and the shell being investigated is of good quality, the model predicts that the change to the new shell will occur sooner than if the animal initially had a low motivation to change (point B, Fig. 5.4). If, however, the shell being investigated is of low quality the decision to reject the shell will take longer if the initial motivation was high (point A) than if the initial motivation was low (point B, Fig. 5.5).

A shell may be of poor quality for a variety of reasons. A shell with two or more negative attributes (e.g. too small and aperture blocked) would be likely to be perceived sooner as being of low quality than would a shell with one negative attribute (e.g. optimal size but with its aperture blocked), even though both shells would eventually be rejected. The speed with which the level of causal factors changes is predicted to depend on the relative qualities of the shell in possession and that being investigated. Thus for any single starting point there may be different speeds with which a shell is rejected (Fig. 5.6) or accepted.

If relatively good-quality shells that are being investigated cause an upward movement of the trajectory and relatively poor ones a downward movement, then shells of the same quality as that occupied will elicit neither upward nor downward movement, and investigation will continue. Theoretically a crab could continue investigating a shell indefinitely unless some other causal factor changes and this may occur for a number of

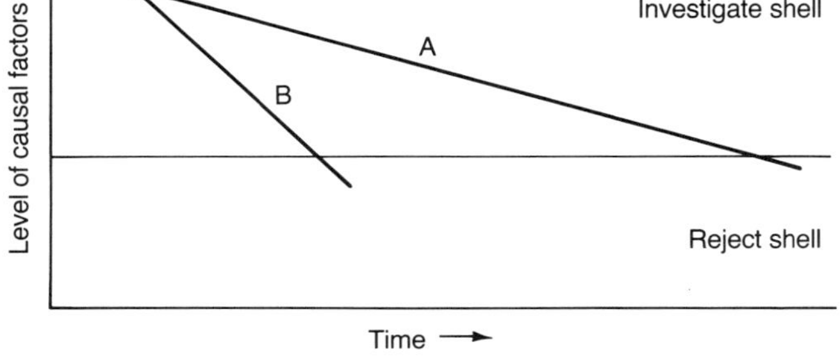

Figure 5.6 A and B represent changes in level of causal factors for crabs encountering poor shells and very poor shells respectively. (From Elwood and Neil 1986a.)

reasons. One reason might be due to increasing distractibility during the course of shell investigation. An example of increasing distractibility was noted by Culshaw and Broom (1980) who found that chicks engaged in either feeding or preening became more distractible during the latter stages of a single bout of either activity. Furthermore, locusts stop walking in response to a lower intensity of sound towards the end of a bout of walking than towards the beginning (Moorhouse *et al.* 1987). That is to say, in each case the animals became increasingly open to the influence of external stimuli. This may be a common feature in animal behaviour and the increasing distractibility may be envisaged as causing a lowering of the causal factors for shell acquisition. In the model presented here this leads to a more rapid movement towards the rejection threshold (Fig. 5.7). Even if a shell is slightly better than that in possession, causing a slow approach of the level of causal factors toward the shell entry threshold, if this approach is sufficiently slow the animal may become distracted and the shell rejected. The increasing distractability of the crabs may thus act as a mechanism not only to prevent the continuation of the activity for too long a time, but also to allow the crabs to test shells by moving into them if there is a low risk but to reject them if a high risk is perceived. The risk may be due to other crabs, which might take the original shell, or to predation.

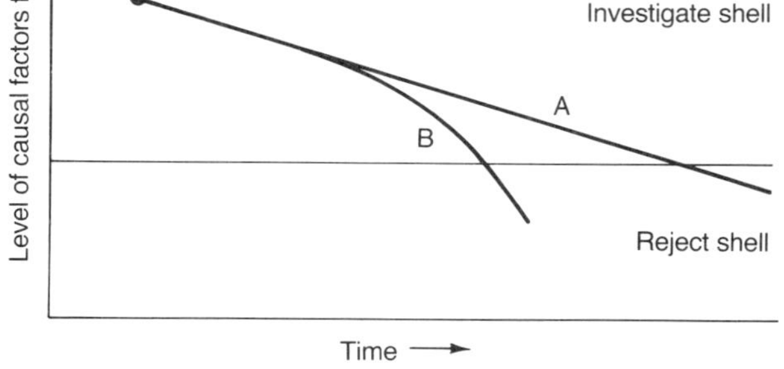

Figure 5.7 Trajectory B illustrates the proposed influence of attention switching during shell investigation. (From Elwood and Neil 1986a.)

TESTING THE MODEL

The model predicts that crabs with a high initial tendency to change shells at the start of investigation (i.e. those in poor-quality shells) would enter the shells more quickly than would crabs with a low initial tendency (Fig. 5.4). It has already been shown that crabs with a high initial tendency change shells more quickly than do those with a low initial tendency (Fig. 5.2). This observation, however, gave rise to the model and therefore cannot be considered a test of it. The model also predicts that crabs with a low initial tendency to change shells (i.e. those in good-quality shells) should reject unsuitable shells more quickly than should those crabs with a high initial tendency (Fig. 5.5). A test of the model was thus made by offering crabs shells with the apertures blocked so that all shells were rejected (Neil and Elwood 1986). The quality of the shells in possession and those being investigated varied among the experimental groups as did certain external stimuli. In this experiment a positive correlation was found between the number of crabs commencing active investigation (a measure of initial level of causal factors for shell change) in each experimental group and the time taken to reject the shells (Fig. 5.8). Thus, crabs with a high initial tendency to change shells do so quickly when a normal shell is offered (Fig. 5.2) and persist with investigation for longer when the aperture is blocked (Fig. 5.8). These data are congruent with the model.

The model predicts that very poor shells will be perceived as being

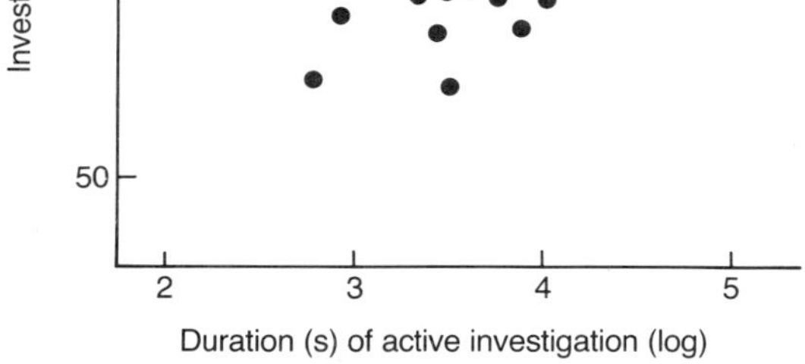

Figure 5.8 A positive correlation was found between the percentage of each of the experimental groups that investigated and the duration of investigation prior to shell rejection. (From Elwood and Neil 1986a.)

unsuitable more quickly than will moderately poor shells (Fig. 5.6). This was also tested by using blocked shells (Neil and Elwood 1986). Crabs were offered either a shell that was much too small (25% of preferred weight) with the aperture blocked or one of the correct size with the aperture blocked. This is a more objective measure of initial causal factors than the use of the proportion of an experimental group initiating an activity. The size of the shells inhabited by the crabs was also varied among experimental groups in a further test of the prediction made in Fig. 5.5. The data obtained (Fig. 5.9) show clearly that both predictions are upheld. Crabs investigating optimal sized shells with apertures blocked persist for considerably longer than do those with very small blocked shells. Furthermore, the worse the shell occupied by the crab the longer the crab persists with blocked shells.

The converse of the prediction of Fig. 5.6 is that excellent shells will be accepted more quickly than will moderate shells if the quality of the shell in possession is kept constant. This has been tested by housing crabs in sub-optimal (50% of the preferred weight) *L. obtusata* shells and then offering either an optimal (100%) *L. obtusata* shell or an optimal size (100%) shell of the less preferred *Gibbula* species. In this case crabs entering the preferred species did so more quickly than crabs entering the less preferred species (Jackson and Elwood 1989b).

The final prediction of the model is that as shell investigation continues

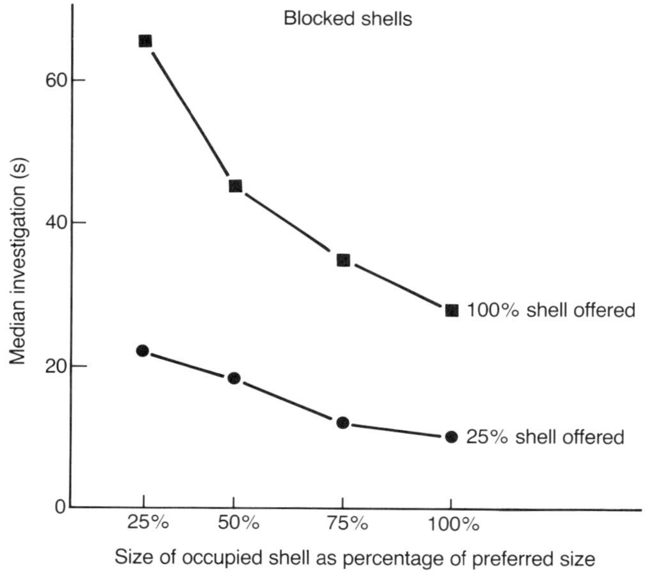

Figure 5.9 Size of shell occupied and of the blocked shell influence persistence times.

crabs should become distractible. This was tested by offering crabs optimal sized shells with blocked apertures and observing whether additional visual stimuli, positioned behind a glass barrier, influenced the time to rejection (Neil and Elwood 1986). The prediction is that the presence of additional stimuli will distract crabs from shell investigation and thus reduce the duration of active investigation. This was found to be true only for crabs housed in very poor shells; the presence of a stimulus (such as a stone, small shell, optimal shell, small, medium or large crab) reduced the active investigation time when compared with crabs without additional stimuli (Fig. 5.10). Furthermore, crabs in moderately poor shells (75%) spent a lower proportion of total contact engaged in active investigation if no additional stimuli were present. Thus when no additional stimuli were present the crab spent much of its time apparently doing nothing whereas those with additional stimuli moved off to investigate those stimuli. These data are thus congruent with the hypothesis that the presence of an additional stimulus distracts crabs during shell investigation.

REFINEMENT OF THE MODEL AND FURTHER TESTS

The data presented above are congruent with the model. Thus it is clear that the timing of decisions about shells is largely determined by the initial motivational tendency to obtain a new shell (determined by the quality of the occupied shell) and the input of information during the assessment of the offered shell. The model developed so far treats shell investigation as a single activity leading to a single decision of whether to accept or reject the offered shell. It was pointed out earlier, however, that shell investigation is composed of several activities, including the two major activities of external

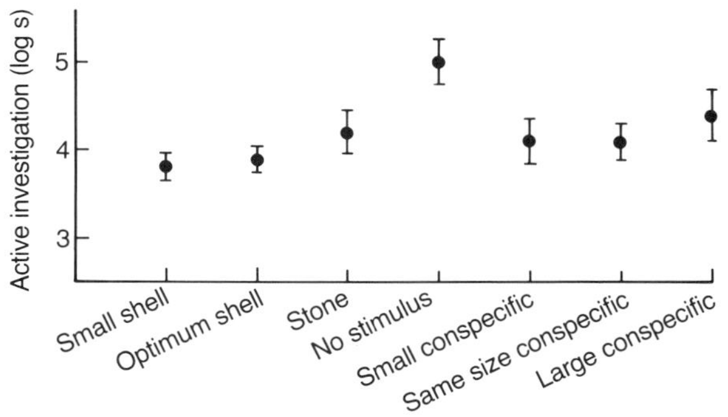

Figure 5.10 Duration of active investigation of crabs in poor shells (25%) is longer when no additional stimulus is present (Group 4) (mean ± SE). Groups 1–3 and 5–7 represent different stimuli presented behind a clear partition.

investigation and aperture investigation. Clearly, a decision is required to switch from one to the other and the model might also apply for each decision in a chain of investigatory activities. To test whether this is the case requires that information obtained from the external investigation and from internal investigation be varied independently of each other.

This has been tested by placing crabs in *L. obtusata* shells that were 50% of the preferred weight so that the initial motivational states of all the crabs were similar (Jackson and Elwood 1989b). These crabs were placed into one of three experiments, each consisting of two groups. In each experiment either *L. obtusata* or *Gibbula* shells of the optimum weight were offered for investigation. Recall from Chapter 3 that the former are greatly preferred over the latter. In experiment 1 the shells were in perfect condition so that both the exterior and the interior differed between the species. In experiment 2 the shells offered had their apertures blocked with dental cement, leaving a smooth surface. Thus only the external features of the shells differed. In experiment 3 the exteriors of the shells were altered by the addition of dental cement, to make the shape of both species like an intermediate between the

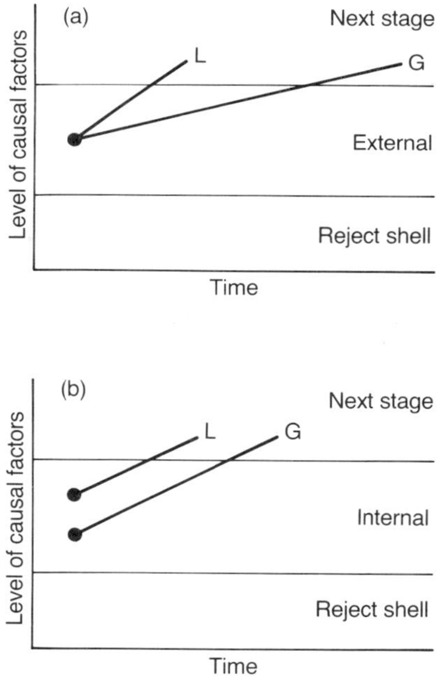

Figure 5.11 Representation of the predicted results of experiment 1: the external investigation above (a) and internal investigation below (b). L = crabs investigating *L. obtusata* shells and G = crabs investigating *G. cineraria* shells. (From Jackson 1988.)

two. The two species thus differed in respect of their apertures alone.

Apart from testing the refinement of the model this procedure should also indicate which sources of information crabs use in their assessment of shells. The refined model predicts the following:

1. If information is gained during external investigation then if unmodified shells of the two species are being investigated the investigation of the external preferred species will be shorter (Fig. 5.11a and 5.12a).
2. When the decision is made to switch activities to aperture investigation the crab starts to turn the shell so that the aperture is uppermost. We presume that information is still being gathered during this process so that, at the commencement of aperture investigation of unmodified shells, those crabs investigating the preferred species will have a higher set of causal factors for shell acquisition than will those investigating the less preferred species (Fig. 5.11b). Thus aperture investigation of preferred species will be shorter because of the high initial level of causal factors and the information gained during aperture investigation.
3. We predict that information is gathered during external investigation and

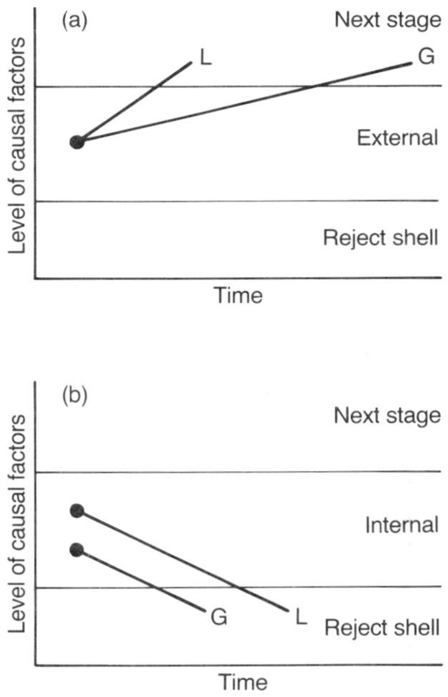

Figure 5.12 Representation of the predicted results of experiment 2: the external investigation above (a) and internal investigation below (b). L = crabs investigating *L. obtusata* shells and G = crabs investigating *G. cineraria* shell. (From Jackson 1988.)

the causal factors are thus higher at the start of aperture investigation for crabs investigating good shells (point 2 above). Thus if the aperture is blocked it is predicted that aperture investigation will continue for longer prior to rejection of the preferred species than for the unpreferred species (Fig. 5.12b).

4. If the exteriors of the shells are modified so that the two species are similar (experiment 3) then there should be no difference in the duration of external investigation (Fig. 5.13a).
5. If point 4 is true then at the start of aperture investigation the level of causal factors for crabs investigating the two species should be the same (Fig. 5.13b). The durations of aperture investigation of modified shells may therefore be compared with those of crabs investigating normal shells (experiment 1). The precise effect of exterior modification should depend upon whether or not the modified shells are perceived as being 'better' or 'worse' than the normal shells, this being indicated by the duration of external investigation.
6. Once a shell change has occurred those crabs that moved to the *Gibbula*

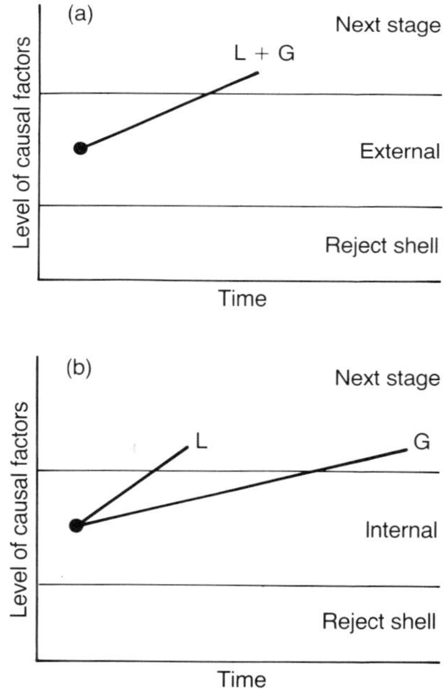

Figure 5.13 Representation of the predicted results of experiment 3: the external investigation above (a) and internal investigation below (b). L = crabs investigating *L. obtusata* shells and G = crabs investigating *G. cineraria* shell. (From Jackson 1988.)

shells should be more likely to reinvestigate the old shell and to move back to that shell.

The results of Jackson and Elwood (1989b) upheld these predictions in virtually every respect. In experiments 1 and 2, when shells with unaltered exteriors were offered, those crabs investigating *Gibbula* shells investigated the exteriors for longer prior to turning the shell than did those with *L. obtusata* shells (Fig. 5.14). Furthermore, in experiment 1 the investigation of the interior of the shell was of a longer duration prior to moving in if the shell was of the unpreferred *Gibbula* species (Fig. 5.15), confirming prediction 2.

The third prediction was that a crab will persevere longer with, i.e. take longer to reject, a preferred species of shell than it will with a non-preferred shell even if both are ultimately found to be unsuitable. In experiment 2, crabs spent longer investigating the blocked apertures of *L. obtusata* shells than they did those of *Gibbula* shells even though they were not gaining any further information about shell structure (Fig. 5.16). In this case the crab often turned the shell again apparently searching for the aperture. Thus, the information gained from the investigation of the exterior of the shell must have had an effect on subsequent aperture investigation and the result may be accounted for by a higher initial level of causal factors at the onset of aperture investigation (Fig. 5.12b).

This, however, raises a problem with the interpretation of the results of

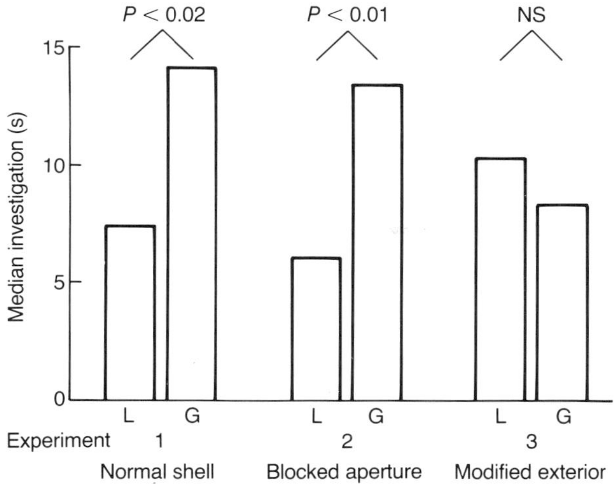

Figure 5.14 The median duration of investigation of the exterior of the shells is shown for the three experiments. L = *Littorina* shells being investigated, G = *Gibbula* shells. (From Jackson and Elwood 1989b.)

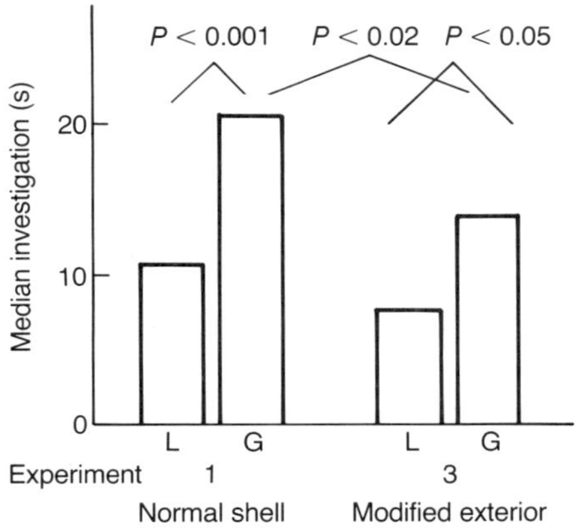

Figure 5.15 The median duration of investigation of the interior of shells is shown for experiments 1 and 3. L = *Littorina* shells being investigated, G = *Gibbula* shells. (From Jackson and Elwood 1989b.)

experiment 1. The difference between the duration of the internal investigations of normal *L. obtusata* and *Gibbula* shells does not necessarily imply that information was being gained from this source. A difference would occur as a result of the external investigation anyway since at this stage the crabs had already gained information indicating that the *Gibbula* shells were of the less preferred species. Thus the level of causal factors would have further to rise to reach the threshold for moving into the shell with *Gibbula* shells (Fig. 5.11b).

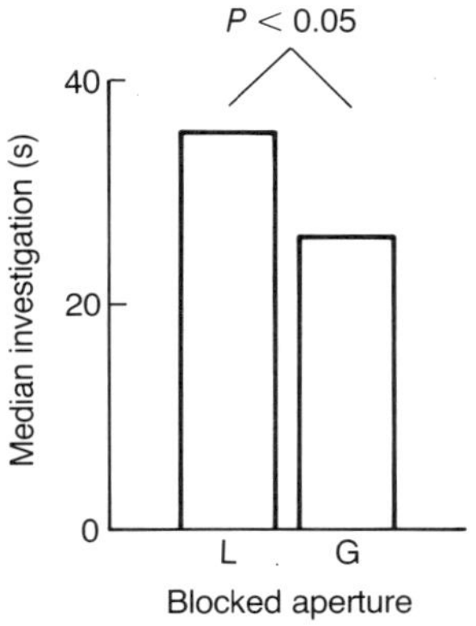

Figure 5.16 The median duration of investigation of the blocked aperture and of subsequent turning of the shell is shown for experiment 2. L = *Littorina* shells being investigated, G = *Gibbula* shells. (From Jackson and Elwood 1989b.)

Experiment 3 was designed to solve this problem by producing shells with identical exteriors so that the level of causal factors would be the same at the start of internal investigation for crabs with either species. As there was no difference between the external investigations of the two species of shell after modification (Fig. 5.14) the crabs appeared to be gaining the same information and hence setting the same level of causal factors at the start of internal investigation (Fig. 5.13b). There was a significant reduction in the external investigation of the modified *Gibbula* shells compared to those of experiment 1 which were unmodified. The increase in duration of external investigation of modified compared to the unmodified *L. obtusata* shells (experiment 1), however, was not significant. The finding in experiment 3 that the investigation of the interior of the *Gibbula* shells remained greater than that of *L. obtusata* shells (Fig. 5.15) shows that information about shell structure was gained from the interior. The data also suggest that the crabs that investigated the exteriorly modified *Gibbula* shells started the interior investigation with a higher level of causal factors than did those with normal shells. This resulted in a shorter investigation of the apertures of the modified *Gibbula* shells (experiment 3) as compared to the unmodified *Gibbula* shells (experiment 1) (Fig. 5.15; prediction 5). There was, however, no significant difference between the interior investigation of modified and unmodified *L. obtusata* shells. These data suggest that the exteriorly modified *Gibbula* shells were perceived by the crabs as being of better quality than were normal *Gibbula* shells. This perception was made during exterior investigation but continued to influence the decision-making process during aperture investigation.

Finally, prediction 6 suggests that having entered the unpreferred species there should be a higher probability of investigating the original shell, i.e. there should be a greater degree of 'satisfaction' with the preferred species. This was confirmed in experiment 1, in which 24 out of 31 crabs in the *Gibbula* group investigated the original shell after moving compared with 17 out of 35 in the *Littorina* group ($\chi^2 = 4.65$, $P < 0.05$) and experiment 3 with 29 out of 34 in the *Gibbula* compared with 15 out of 35 in the *Littorina* group investigating the original shell ($\chi^2 = 11.67$, $P < 0.001$). Furthermore, of those crabs that changed shells in experiments 1 and 3, more crabs returned to their original shells from their new *Gibbula* shells (12 of 31 and 12 of 34 respectively) than did those with new *L. obtusata* shells (0 of 35 in both experiments). These differences are significant for experiment 1 ($\chi^2 = 14.06$, $P < 0.001$) and experiment 3 ($\chi^2 = 12.59$, $P < 0.001$) and once again confirm that *L. obtusata* is the preferred species.

It is clear from the above data that the idea of each stage of investigation being a 'fixed action pattern' is too simple (Reese 1963). Each stage varies in duration and, as noted in Chapter 4, in the appendages employed. It thus seems more reasonable to view each stage as a process of information gathering that results in a motivational change, with the decision of whether to continue to the next stage depending on the nature of that motivational change. Although the data reported so far are compatible with the model, it might be tested further by probing the underlying motivational state by use of an interruption technique (Jackson and Elwood 1990). This should provide an independent method of examining how information alters motivational state.

INTERRUPTIONS OF ONGOING BEHAVIOUR

Differences in the underlying motivational state may be assessed by exposing animals to a stimulus that may alter the ongoing activity. This has been achieved with chicks (Culshaw and Broom 1980) and with locusts (Moorhouse *et al.* 1987) with respect to changes in motivational state during bouts of a particular activity. Both chicks and locusts were more likely to be startled by a novel stimulus if the stimulus occurred towards the end of a bout and this was interpreted in terms of changes in attention or of positive feedback (lock-on) mechanisms in the early stages of a bout (see Chapter 1). In a slightly different context, sticklebacks were less likely to respond to a model of a predator when feeding either at a high rate or on high-density swarms (Milinski 1984). Again, differences in the attention systems were postulated to account for this behaviour. Exposure to model predators also influenced subsequent foraging tactics in fish (Milinski and Heller 1978; Metcalf *et al.* 1987).

The underlying motivational state may vary during a bout of activity for a number of reasons associated with causal factors relating to that activity as well as those relating to other activities (Chapter 1). One such factor, which has received little study to date, is that the animal may be gathering information that directly influences its motivational state. Thus, in the case discussed above, it was demonstrated that information gathered during external investigation influences the motivational state at the start and during the subsequent act of aperture investigation (Jackson and Elwood 1989b). If, indeed, the motivational state is altered by information then this might be demonstrated further by using an interruption technique. In particular, it might be predicted that when a crab perceives that the shell being investigated is of low quality it should be more easily distracted than if the shell is perceived as of high quality. In this manner the interruption technique has been employed to discover the stage at which the quality of a shell is discriminated (Jackson and Elwood 1990).

Crabs were given *L. obtusata* shells 75% of their preferred weight. The following day they were each offered a single *L. obtusata* shell either of 100% or of 50% of the preferred weight, i.e. a potential gain or loss respectively. During the subsequent shell investigation a novel stimulus was presented.

This was a black card moved over the crab at a height of 10 cm causing a shadow to fall on the crab for about 1 second. This stimulus was employed because crabs had been noted to withdraw into their shells in response to overhead movement in field conditions. The stimulus was presented at one of three points in shell investigation: (i) immediately after the crab had started to hold the shell, i.e. before external investigation, (ii) immediately after it had turned the shell over, i.e. after external investigation, and (iii) after five seconds of aperture investigation.

Contrary to expectation there was no difference in the probability of a startle response by the crab, involving partial or complete withdrawal into the shell, dependent upon the quality of the shell being investigated regardless of when the stimulus was given. There was, however, a decreased probability of startle when the stimulus was presented immediately after the animal had turned the shell. This might be due either to a high level of motivation at that point, which is, in effect, the very onset of aperture investigation, or alternatively to the attention being directed at the aperture as it is turned into view so that the novel stimulus was less likely to be detected.

The crabs that were stimulated at the earliest point in the investigation should have gathered the least information. In these cases, even though the probability of interruption was the same for the two qualities of shell, those crabs investigating good shells (100%) had interruptions of a significantly shorter duration than did those investigating poor shells (50%) (Fig. 5.17). Thus the underlying motivational state of crabs in the two groups differs just after taking hold of the shells, which indicates that some perception of shell quality has occurred by that stage. A similar difference in the duration of interruption was found when the novel stimulus occurred after five seconds of aperture investigation (i.e. longer interruptions when investigating a poor shell), but not if the interruption occurred at the turning stage. In this latter case interruptions were particularly short (Fig. 5.17).

Because this experiment showed that crabs could perceive some aspects of shell quality by the time of first contact a second experiment was conducted with novel stimuli presented prior to contact with the shells. In this experiment (Jackson and Elwood 1990) the quality of the shell occupied was also varied, some crabs being in *L. obtusata* shells 50% and others 75% of the preferred weight. These crabs were offered either *L. obtusata* or (the less preferred) *G. cineraria* shells of the preferred size with the novel stimulus applied as the crab approached to within 1 cm of the offered shell.

Once again the quality of shell on offer had no effect on the probability of a startle response (most crabs showed some response) and neither did the quality of the occupied shell. There were, however, significant effects of both variables on the duration of the interruptions. Crabs approaching the less preferred *G. cineraria* shells had longer interruptions than those

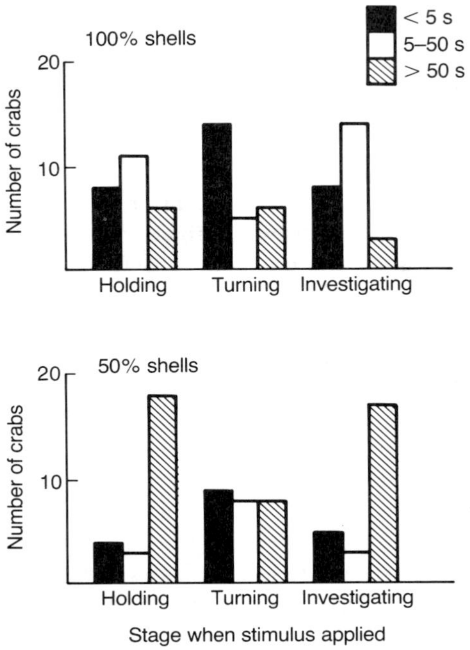

Figure 5.17 The duration of the interruption in assessment behaviour following presentation of a novel stimulus at different stages of the investigation for crabs occupying shells 75% of preferred weight and offered 100% or 50% shells. (From Jackson and Elwood 1990.)

approaching *L. obtusata* shells, indicating that even prior to contact some information had been gained (Fig. 5.18). When approaching *L. obtusata* shells the size of the shell a crab was occupying had no effect on the duration of interruption. When *G. cineraria* shells were being approached, however, crabs in 50% shells had shorter interruptions than did those in 75% shells. Thus, even prior to contact the underlying motivational state is influenced by both the quality of the shell in possession and of that being approached.

Combining the appropriate data from the two experiments allowed examination of the effects of the time that the novel stimulus was applied (Fig. 5.19). This indicated that interruptions occurring prior to contact with the shell were of considerably greater duration than were those that occurred after contact, an effect possibly due to the amount of information the animal had at those times. It is reasonable to assume that even if crabs have some information prior to contact (as shown above) the amount of information is less than that obtained after contact. There may, however, be differences

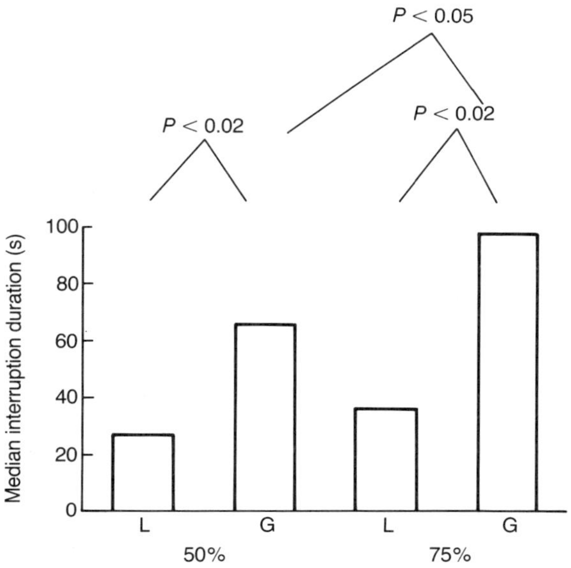

Figure 5.18 The median duration of the interruption in assessment behaviour, following presentation of a novel stimulus, of crabs occupying of 50% or 75% shells and offered shells of high quality (L = 100% *L. obtusata*) or low quality (G = 100% *G. cineraria*). (From Jackson and Elwood 1990.)

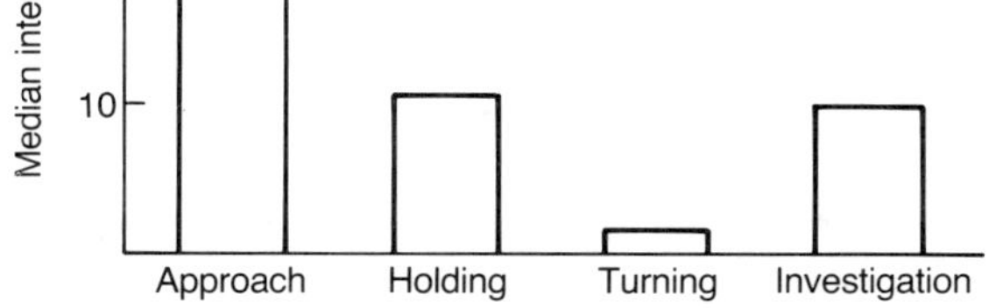

Figure 5.19 The median duration of the interruption in assessment behaviour, following presentation of a novel stimulus, of crabs from both experiments occupying 75% shells and offered 100% *L. obtusata* shells stimulated at different stages of the investigation. (From Jackson and Elwood 1990.)

in attention of these animals whilst engaged in different activities which might, at least in part, account for these differences in duration of interruption.

Crabs exposed to the novel stimulus at the approach, contact and turning stages were compared with respect to their subsequent behaviour when they resumed investigation. Those interrupted prior to contact had the longest subsequent aperture investigations; in other words the longer the crab had to recover from the novel stimulus the longer the duration of aperture investigation. Furthermore, for crabs stimulated during approach there was a significant positive correlation between the duration of the interruption and the duration of the subsequent aperture investigation (Fig. 5.20). This suggests that the more severe the disturbance (judged by duration of interruption), the more cautious the crab was in its subsequent shell changing. This is similar to the change in foraging strategy of salmon parr after exposure to a model predator. Immediately after seeing the 'predator' the fish delayed orientating towards passing food particles and delayed attacking food until the food had reached its closest point. This effect was temporary, however, and the fish soon resumed orientating towards and attacking food at a distance (Metcalf *et al.* 1987).

It is clear from these experiments that information alters the underlying motivational state as judged by the behaviour subsequent to interruption. Crabs occupying low-quality shells or investigating high-quality shells have shorter interruptions. These results therefore support the idea that features of both the shell inhabited and the shell being investigated combine to determine the level of causal factors for shell acquisition (Fig. 5.3).

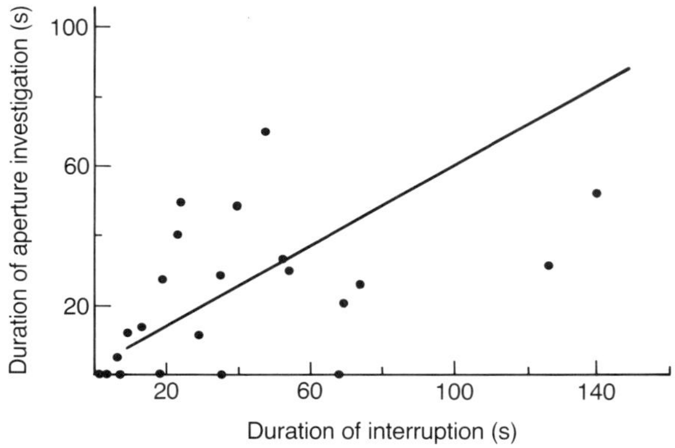

Figure 5.20 The relationship between the duration of the interruption in a crab's investigation and the duration of subsequent internal investigation for crabs interrupted at the approach stage ($r = 0.595$; $P < 0.005$). (From Jackson and Elwood 1990.)

MEMORY FOR SHELLS

The above analyses seem to indicate that information concerning shells is gained prior to contact by visual means, and also during external and then aperture investigation by tactile means. Furthermore, information acquired during one stage influences the decisions made at later stages. We have discussed this in terms of changing levels of causal factors, or motivation, but another way to describe this process is to say that information gained at one stage is remembered at a later stage. In this section we examine the possibility that crabs can retain information from one encounter with a shell to another encounter with that same shell.

It has been shown that crabs may prefer shells that they have previously occupied when given a choice between these shells and novel shells, which may indicate memory of particular shells (Grant 1963). In this study, however, the result may have been due not to memory but to the better fit of previously occupied shells. This problem has been overcome in a recent study in which crabs were offered shells with blocked apertures, which caused their rejection (Jackson and Elwood 1989a). Crabs then had the opportunity to investigate the same shells again and the duration of this second investigation was compared with the first. A reduction in the duration of the second investigation might be due to either fatigue or memory and in the latter case it might be the shell or the location of the shell which is remembered. These possibilities were all examined.

Three groups of crabs were used, all of which were housed in *L. obtusata* shells 50% of the preferred weight. All crabs were then placed with two *L. obtusata* shells of the preferred weight but with blocked apertures, and allowed to complete the investigation of one of these shells. Group 1 crabs were then moved back to the start point in the observation dish, Group 2 crabs were moved back to the start point and the positions of the two shells exchanged, and Group 3 crabs were left undisturbed.

For their next investigation, there was no difference between the groups in the proportion of crabs that reinvestigated the first shell or investigated the novel shell. For those that investigated the novel shell there was no difference between the duration of that investigation and that of the first investigation i.e. there was no sign of fatigue. If, however, the crab reinvestigated the original shell there was a significant decline in duration for this second investigation for all three experimental groups. There were no differences between the three experimental groups and the combined data are shown in Fig. 5.21. These data show that fatigue is not responsible for the decline in duration of investigations in repeat investigations as it did not occur when crabs investigated novel shells. Furthermore, the decline occurred even if the familiar shell was found in a novel location, which demonstrates that it is the shell that is remembered and not its location. This

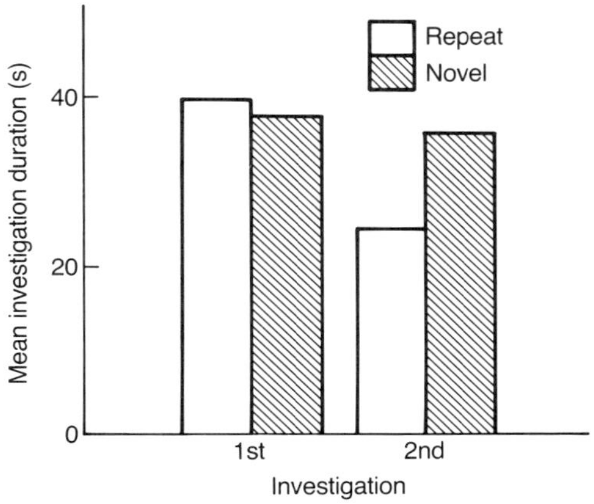

Figure 5.21 The mean durations of the first and second investigations made by both repeat and novel investigators. The decline between first and second investigation of the same shell (repeat) is significant ($P < 0.05$) but that for different shells (novel) is not. (From Jackson and Elwood 1989a.)

memory of shells may decay with time since the longer the interval between the first and second investigations of the same shell the smaller was the decrement in the second investigation (Fig. 5.22) (Jackson and Elwood 1989a).

These data show that information gathered in one investigation is remembered at a subsequent investigation and thus support the hypothesis that information gathered at one point in an investigation may be remembered at a later stage in the same investigation. That the crabs were not chemically

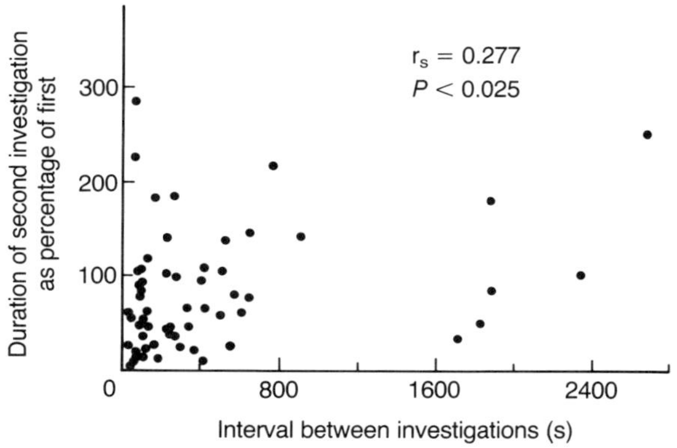

Figure 5.22 The difference between the durations of two investigations of a shell is reduced the greater the time lag between them. (From Jackson and Elwood 1989a.)

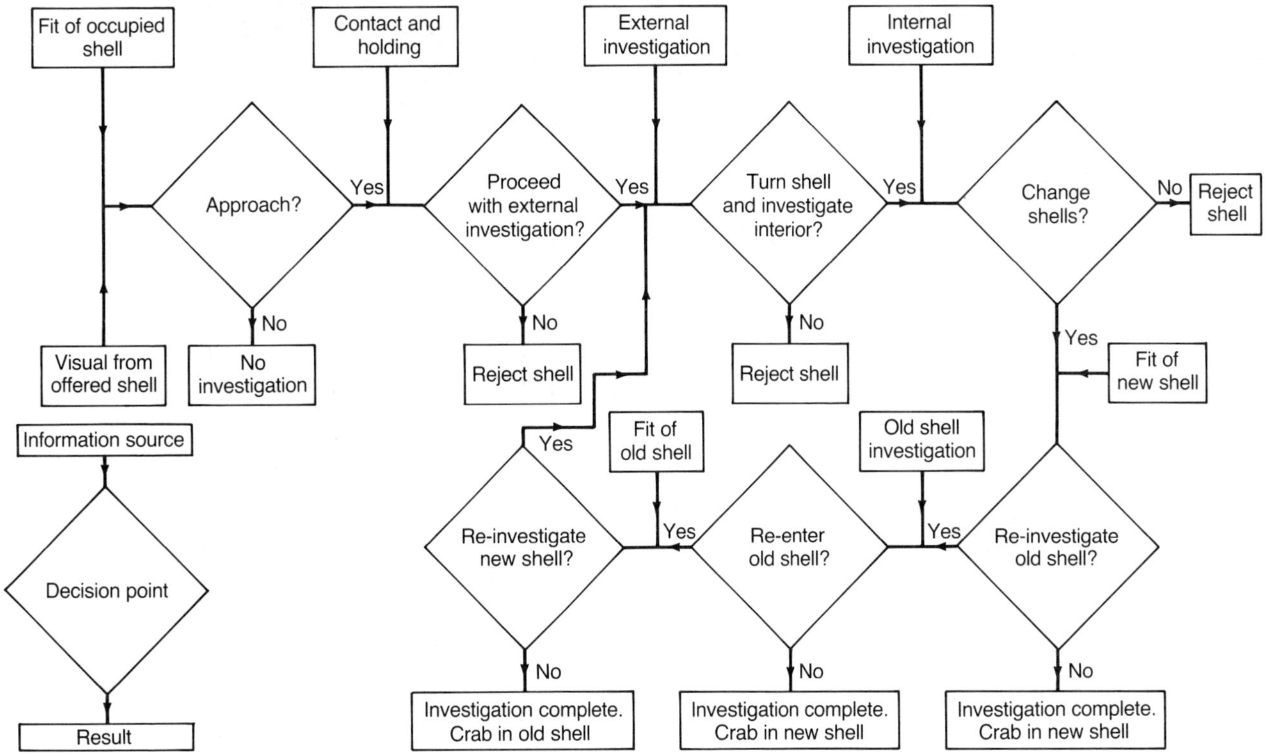

Figure 5.23 The complexity of the complete shell assessment is simply due to the large number of potential decision points. (From Jackson 1988.)

marking the shells was inferred from the finding that antennal contact with novel or familiar shells did not influence the probability of investigation.

CONCLUSIONS AND SUMMARY

Shell assessment is a remarkably complex behaviour involving several sources of information, several sensory systems and several distinct decision points. These are summarized in Fig. 5.23. This figure indicates that the fit of the occupied shell and visual information from the offered shell influence the decision of whether to approach the shell. This information, plus that gained during initial contact and holding, influences the decision of whether to investigate the exterior. This information is then added to that gained during external investigation and the total influences whether the interior is investigated and so on. Clearly, information gained at one stage is available at later stages, i.e. information is remembered. The entire process may be viewed as a series of relatively simple decision points, with each decision-making process depending on rising or falling levels of causal factors that cross thresholds for either continuing to the next stage or for rejection. The complexity arises because of the number of such decision points in any one shell assessment process. Under normal circumstances there are at least five such decision points: (i) to approach, (ii) to investigate the exterior, (iii) to investigate the interior, (iv) to change shells, and (v) to re-investigate the old shell. There may, however, be many more decision points because the crab may change back to the old shell and start investigating the new shell again.

This detailed examination of an assessment process has demonstrated the importance of information gathering in decision-making processes. The motivational model may be used in studies of other assessment processes that involve acceptance or rejection of a resource. It should be applicable to a wide variety of situations including decisions about habitat choice, nest-site selection, foraging, mate choice etc. However, the model is limited to discrete two-way decisions such as accept or reject. Decisions such as how many eggs a parasitic hymenopteran should lay (Chapter 1) or decisions concerning the duration of particular activities will require an alternative approach. Decisions such as these, with many possible outcomes, may be examined using hermit crabs in agonistic situations and are discussed in the final chapters of this book.

CHAPTER 6

Agonistic behaviour

Most ecological studies of competition amongst hermit crabs have concentrated on the hypothesis that shells are a limiting resource (Vance 1972a; Bertness 1980, 1981a, b; Abrams 1981a, b, c). Competitive exclusion may occur because dominant species are able to exclude subordinate species from particular species of shell (Abrams 1980, Bertness 1981a). Both inter-and intraspecific competition for shells is mediated primarily by agonistic interactions called 'shell fights' at the end of which one crab might evict the other and make its choice of shells (Reese 1963; Hazlett 1966b, d). The central importance of shells as an essential resource to most hermit crab species is well documented (Chapter 2) and, because shells are easily quantified, and the mechanisms of competition well described, ecological studies of this group have virtually excluded consideration of all other resources. However, both inter- and intraspecific competition occur for food and space and intraspecific competition occurs for access to mates (Reese 1962b, 1983; Helfman 1977a). Indeed, hermit crabs show a variety of displays, other than shell fights, by which such competition may be mediated and it is these displays that are considered in the present section; shell fights are discussed in subsequent sections. The two types of interaction are not entirely separate since shell fights are usually preceded by displays but not all displays are followed by fights.

DISPLAYS

Some studies of the displays of hermit crabs are purely descriptive (e.g. Reese 1962b; Hazlett 1966a,d, 1969a; Helfman 1977a) whereas others are experimental and have investigated the communicatory function of the displays by using models (Hazlett 1968a, 1972c), modified crabs (Dunham 1978a) or analysis of interactions between normal crabs (Hazlett and Bossert 1965, 1966; Hazlett and Estabrook 1974a, b; Hazlett 1980b). Other studies have examined various factors that elicit displays (Hazlett 1966b, 1970a, c) and the role of experience in agonistic encounters (Hazlett and Provenzano 1965; Hazlett 1969b; Courchesne and Barlow 1971).

There is a great similarity in displays throughout the two major families of hermit crabs (Hazlett 1966d). The chelipeds and/or ambulatory legs are used and in many species these are distinctively coloured (Hazlett 1966d; Dunham 1978b). The three most common displays are:

1. *Cheliped presentation* – the major (often with the minor) cheliped is held forwards but with the chela(e) approximately perpendicular to the substrate (Fig. 6.1).
2. *Cheliped extension* – the major (often with the minor) cheliped is moved forwards with the pincer approximately horizontal to the substrate (Fig. 6.2). The chela(e) may be open. The movement may be part of the display but may also result in the opponent being struck by the cheliped(s) (Reese 1962b).
3. *Ambulatory raise* – at least one of the ambulatory appendages is extended sideways away from the body and held above the substrate (Fig. 6.3).

During the cheliped displays the ambulatory appendages may be spread apart slightly, giving the animal a more secure base, or the crab may push up on the ambulatories so that it appears higher (Fig. 6.4) (Hazlett 1968a). On other occasions the ambulatories may be held so that the crab appears lower than normal (Hazlett 1968a).

The functions of these displays have been investigated in *P. bernhardus*

Figure 6.1 Cheliped presentation by both crabs, chelae perpendicular to substrate.

Figure 6.2 Cheliped extension by both crabs, chelae horizontal to substrate.

Figure 6.3 Ambulatory raise by both crabs.

Figure 6.4 High posture approach by both crabs.

with the use of models (Hazlett 1968a, 1969a). The models were constructed from dried exoskeletons, the joints of which were glued to hold the limbs in the desired positions. Each model was then glued to the end of a glass rod and presented to crabs to assess their responses. The models were in (i) neutral position, (ii) cheliped presentation, (iii) cheliped extension and (iv) ambulatory raise. Further models assessed the effect of high or low stance either with or without cheliped presentation.

Ambulatory raise has been described as rare in *P. bernhardus* and the model of this position had similar effects to that of the neutral position (Hazlett 1968a, 1969a). Observations of small crabs in our laboratory, however, suggest that it may be more common than previously suggested (pers. obs.). Models of both types of cheliped display were likely to elicit a response, most commonly retreat (either moving away from the model or withdrawing into the shell). Crabs were likely to display with their chelipeds either as they maintained their position or retreated. There was, however, no apparent difference in the responses of crabs to the models of the two types of cheliped display. A model in the low posture without any cheliped presentation was more likely to be ignored than one in the neutral position. Low posture also appeared to cancel the effects of a cheliped presentation. High posture, however, was more likely to elicit a retreat than was the neutral posture and high posture combined with cheliped presentation was particularly likely to elicit a cheliped display. It seems that the high posture enhances the apparent size of the crab and in another species (*Calcinus tibicen*) just being on a flat stone 6 mm thick gave crabs a significant advantage in forcing opponents to retreat (Hazlett 1966b).

The colour and size of the chelipeds influences the effectiveness of the cheliped presentation display in *P. bernhardus* as demonstrated by the use of models (Hazlett 1969a). *P. bernhardus* has a reddish stripe along the chelipeds which are otherwise tan coloured. When this stripe was eliminated, leaving the chelipeds uniformly tan, the cheliped presentation display was more likely to be ignored. If the chelipeds were uniformly reddish-brown then a response was even more likely than if the model was of a normal colour (the most common response to these models being some form of display). In *Calcinus laevimanus* the white patch on the manus of the cheliped was shown to be important in determining the winner of agonistic interactions and crabs that had the patch painted out had reduced success (Dunham 1978b). Models of *P. bernhardus* in which the manus of the large cheliped was from a much larger crab caused retreat and withdrawal into the shell whereas a model with a small manus was likely to be ignored (Hazlett 1969a). This indicates the importance of the size of the cheliped in these displays and in settling disputes in general, and concurs with suggestions that cheliped size determines which crab will attack in a shell fight (Neil 1985). Hazlett (1968a), however, found no clear correlation between the size of the animal being tested and its reactions to a standard model. This is unusual, as studies on a variety of other species have consistently found that relative size is an important factor in causing the smaller crab to retreat (Dunham 1978b; Hazlett 1966b, 1968f). A major function of these display components thus seems to be to aid assessment of relative size and thus to determine access to a particular resource.

A problem with the use of models is that they provide only a static visual image and therefore much of the information that may be inherent in the movement of the normal display is lost (Hazlett 1968a). Observations in the field suggest that the cheliped extension display is the one most likely to produce a response from an opponent and yet the cheliped presentation display is the most effective when models are used (Hazlett 1968a). This is probably because the presentation is a relatively static display and hence well represented by the model, whereas the effectiveness of cheliped extension is due largely to the extension movement that is missing in the model.

An alternative approach to the study of the function of displays, which overcomes problems inherent in the use of static models, is to examine the sequence of interactions between two normal animals (Hazlett and Bossert 1965, 1966; Dunham and Tierney 1983). Observations of a variety of species, including *P. bernhardus* (Hazlett 1982b), demonstrate that certain acts by one animal may change the probability of occurrence of particular responses in the other animal, and these changes in probability enable the information content of the display to be calculated. The application of 'information theory' makes it possible to investigate two aspects: first, the importance of an animal's own previous behaviour in determining its subsequent activities and second, the influence of signals on the behaviour of a recipient. Information theory has been especially useful in studies of agonistic interactions of crustaceans (Hazlett and Bossert 1965; Dingle 1969; Hazlett and Estabrook 1974a, b; Rubenstein and Hazlett 1974; Hyatt 1983). Of particular interest was the finding that early in a display sequence of *Calcinus tibicen* the information content per display unit was low but increased during a sequence; that is, the replies became more predictable (Hazlett 1980b). Furthermore, the eventual winners appeared to act as if they obtained more information from losers than losers did from winners. Eventual winners modified their behaviour more predictably in relation to the eventual loser's acts than vice versa (Hazlett 1980b). This, however, leaves the question as to how the winner is determined. Is the winner the crab that alters its behaviour relative to the other crab or does one crab lose because it behaves less predictably or more variably? It is clear that many important questions remain concerning the use of particular display components in this group.

Hermit crabs may also be responsive to water movements caused by the opponent during agonistic encounters (Barron and Hazlett 1989). *Calcinus tibicen* were observed to generate directed water currents, particularly in association with an ambulatory raise, and these water currents usually caused the opponent to turn away or retreat. They were particularly common between animals that were similar in size and shell quality and seemed to be effective in determining the winner in these encounters.

During an agonistic encounter it is not only the movement and posture of the crabs that influences the outcome, but the shells they inhabit also have an effect (Hazlett 1966b, 1970a). For example, the orientation of an empty shell influenced whether a crab approached it or retreated (Hazlett 1966b). Crabs tended to retreat from the front of the shell but approach towards the back. Clearly, approaching from the front poses more risks of counter attack should the shell be inhabited. Increasing the apparent size of an animal's shell by the addition of plastic increased the probability of an opponent retreating (Hazlett 1970a), suggesting that one aspect of the crab's assessment of the size of the opponent is by the size of the shell.

Isolation of crabs increases the probability of their initiating an interaction and of their use of particular display components (Hazlett 1966b; Courchesne and Barlow 1971). Group housing, on the other hand, results in a decrease in agonistic encounters within the group, probably due to individual recognition and memory of particular outcomes of interactions (Hazlett 1968c, 1969b). Crabs may also be influenced by having either won or lost encounters in the recent past irrespective of memory of the individuals involved. If housed with either larger or smaller individuals their subsequent success when tested with strangers is markedly influenced by whether they have recently won or lost encounters (Hazlett 1966b). Finally, a period of food deprivation results in increased aggressiveness and heightened probability of winning an encounter (Hazlett 1966b) as also occurs in other crustaceans (Hazlett and Estabrook 1974a; Hazlett *et al.* 1975).

SHELL FIGHTS

As suggested by the name, 'shell fights' are agonistic interactions during which ownership of shells may be changed. Shell fights are usually preceded by displays (previous section) that on other occasions do not lead to a fight so the start of a shell fight is defined as when one animal withdraws into its shell and that shell is grasped by the other animal. The animal that withdraws into its shell is referred to as the 'defender' and that which moves forward and grasps the defender's shell is called the 'attacker'. The first question that arises is what factors cause a crab to withdraw into its shell rather than display, ignore or retreat?

Hazlett and Bossert (1965) provide data on intraspecific interactions of three tropical hermit crabs (*Pylopagurus operculatus, Clibanarius tricolor* and *Calcinus tibicen*). Three acts on the part of one crab can result in withdrawal into the shell by the other. These are approach, ambulatory raise and cheliped display. Of these it is some form of cheliped display that is most likely to result in withdrawal for all species. Approach and ambulatory raise by one crab are more likely to result in retreat or display by the other.

A study of interactions that precede the shell fights of *P. bernhardus* shows that larger crabs are more likely to approach a smaller crab rather than vice versa (362 compared to 143) (Elwood and Glass unpublished). Some of these approaches are accompanied by cheliped displays on the part of the approaching animal (20% for larger and 35% for smaller crab), but only cheliped displays by a larger crab result in the other crab withdrawing into its shell (Fig. 6.5). The most common result of approach, however, is that antennal contact occurs, with both crabs (usually) touching the other with their antennae. This is followed by the larger of the two (very rarely the smaller) moving forward rapidly and grasping the shell of the other crab to initiate a shell fight (Figs 6.6 and 6.7). In some cases the animals grapple with their chelipeds for some seconds prior to one animal withdrawing into its shell, this being more common if both animals are large and the size difference small.

Some shell fights start without an approach towards the front of the other crab and we refer to these as 'sneaky attacks'. One crab may approach from the rear and grasp the opponent's shell causing the latter to withdraw into its shell. In crowded conditions one crab may retreat from one interaction

Figure 6.5 The larger crab in a *G. cineraria* shell approaches a crab in a *L. obtusata* shell to initiate a shell fight.

Figure 6.6 The larger crab makes contact with the shell of the defender.

Figure 6.7 The defender partially withdraws as the attacker grasps the defender's shell.

only to back into another crab and be forced to withdraw into its shell (Hazlett 1968c). Occasionally in our experiments, after crabs are placed in an observation tank, one crab may emerge to normal position from within its shell more quickly than its opponent and approach whilst the opponent is still withdrawn. Again, this results in the attacker grasping the defender's shell without initial displays.

The following description of the course of shell fights is taken largely from an extensive study of some 500 staged fights of *P. bernhardus* from our study site (Dowds 1983). The attacker, whilst grasping the defender's shell with its legs, moves its chelipeds and/or legs over the external features of the shell (Fig. 6.8). The attacker may also climb on the back of the defender's shell so that its own shell is no longer in contact with the substrate (Fig. 6.9). Hazlett (1966d, 1968b, 1970a) describes a behaviour exhibited by the defending crab in these circumstances which he calls 'dislodgement shaking'. The defender moves its shell rapidly up and down, often shaking off the attacker. This behaviour was not observed in our studies but this may be due to the differences in geographical location of the crabs (Sweden and N. Ireland) and/or size and age differences between our smaller and Hazlett's larger individuals.

After external investigation, the defender's shell is turned by the attacker

Figure 6.8 The attacker feels over the exterior of the defender's shell.

Figure 6.9 The attacker climbs over the defender's shell.

so that the aperture is uppermost and the shell is gripped by the attacker's walking legs (Fig. 6.10). Meanwhile the attacker inserts one or both chelipeds into the aperture of the defender's shell, sometimes along with one walking leg. This period of apparent aperture exploration continues for some time and is interspersed with bouts of shell rocking and shell rapping. Shell rocking usually precedes shell rapping and is similar to the 'positioning movements' described for older, larger specimens of *P. bernhardus* (Hazlett 1967, 1970a) and other pagurids (Hazlett 1966d). The defender's shell is moved around its columellar axis a few times and then shell rapping commences. Whereas rocking is gentle and occurs only once or a few times in succession, rapping is much more vigorous and occurs in bouts with each rap measurable as a discrete unit. The attacker holds the defender's shell steady with its walking legs and hits its shell against the defender's shell, or pulls the defender's shell against its own. Bouts usually consist of three to six raps and are interspersed with periods of aperture exploration or turning of the shell. Hazlett (1966d) reports a behaviour known as spasmodic shaking which occurs prior to rapping in some Carribean species (*Pagurus miamensis, P. pygmaeus, P. bonairensis* and *P. marshi*); however this activity was not observed in our studies.

From the start of a shell fight until the period of rapping ends, the

Figure 6.10 After turning the defender's shell over the attacker raps its shell onto that of the defender and attempts to grasp the defender's appendages.

defending crab remains tightly withdrawn into its shell, although it sometimes shows a defensive behaviour known as cheliped flicking. This behaviour involves the defender moving one or both chelipeds through an arc in the aperture of the shell and, according to Hazlett (1970c), encourages the attacker to release its hold on the defender's shell. Cheliped flicking may occur during aperture exploration, rocking, rapping or all three.

At almost any time during the above activities the attacker may grasp one or more of the defender's limbs and pull vigorously, sometimes pulling the defender from its shell (Fig. 6.11). This use of force has been observed by Allee and Douglis (1945), Orians and King (1964), Elwood and Glass (1981) and Dowds (1983) but not by Hazlett (1966d, 1970a, c) who states that no physical contact between the crabs occurs. Whilst differences in the behaviour of *P. bernhardus* (Hazlett pers. comm.) and other crabs have been attributed to age and experience (Hazlett and Provenzano 1965), force between larger older individuals has been observed in our laboratory. In our observations one crab rarely evicts another without some physical contact and our recent observations of slow-motion video recordings indicates that the defender signals its readiness to leave its shell by rapidly tapping the attacker's chelipeds with its (defender's) walking legs. Invariably, however, the defender is 'assisted' by the attacker holding the defender in its chelipeds (Fig. 6.12). This also occurs in *P. gaminus* (Imafuku 1989).

Figure 6.11 The defender is partially pulled out of its shell.

Figure 6.12 The defender is pulled out of its shell by the attacker and held aloft for a short period.

After eviction, the winner explores both the external and internal architecture of the now vacant shell using chelipeds and walking legs (Fig. 6.13). At this stage the naked loser may still be gripped by the attacker and held away from the shell or it may be released. If the latter, it may approach the empty shell but is driven away by the victor rapidly extending its chelipeds. Subsequent to this period of shell investigation, the winner may or may not enter the empty shell. If it does change shells (Fig. 6.14), as did the vast majority of victorious crabs in our studies, then it normally engages in an investigation of its old, now empty shell (Fig. 6.15). The victor also examines the new shell by feeling back with the walking legs, withdrawing into the new shell several times and raising the shell up off the substrate. These are the same activities seen in normal shell investigation (Chapter 4). Eventually the victorious crab either moves off in its new shell or returns to its old shell. Throughout this period the empty shell is held by the victor and only when the winner moves away in its chosen shell is the naked loser free to enter the discarded one. It does this usually without prior examination. As the crabs retreat from the interaction site, they may exhibit a cheliped extension and/or stand on the tips of their chelipeds and ambulatories (Fig. 6.16).

If eviction of the defender occurs we refer to the fight as a 'complete fight' (regardless of the final shell choice of the victor). If no eviction occurs, the

Figure 6.13 The attacker examines the defender's shell while the evicted crab waits nearby.

Figure 6.14 The attacker changes to the preferred *L. obtusata* shell.

Figure 6.15 The attacker examines its old shell.

Figure 6.16 The attacker retreats from the encounter site in its new shell whilst displaying with a walking leg and assuming a high posture.

fight is 'incomplete' (Dowds and Elwood 1983). In an incomplete fight the attacker stops fighting and moves away, and this may occur at any time during the course of a shell fight. In particular, the attacker may stop before it commences shell rapping.

Dowds (1983) analysed 91 complete fights in detail, first by producing a correlation matrix for the durations of 18 activities performed during shell fights and then subjecting the data to a principal components analysis. These fights were divided into three sections: first, the period up to eviction of the defender (pre-eviction stage), second, the period prior to the victor changing to the defender's shell (post eviction/prechange stage) and finally, activities that occur after shell change (post-change stage). Dowds (1983) recorded six, three and seven different activities pertaining to the three periods, respectively, for the attacker and one for the defender as well as total fight time up to eviction (Table 6.1). Four of the six pre-eviction activities of the attacker show positive correlations with each other. These are: shell investigation, shell rocking, shell rapping and grabbing of the defender. In the post-eviction/prechange stage of the fight a positive correlation occurs between the investigation of the now-empty shell and grabbing at the evicted loser of the fight. In the post-change stage there are a number of positive correlations

Table 6.1. List of shell fight variables examined by Dowds (1983)

1. Grab crab in pre-eviction stage – one crab holds any portion of the other crab with either or both chelipeds
2. Grasp shell in pre-eviction stage – one crab holds the other's shell with either or both chelipeds and walking legs
3. Climb on shell in pre-eviction stage – one crab climbs onto the other crab's shell so that its own shell is no longer on the substrate
4. Shell investigation in pre-eviction stage – one crab investigates the exterior and/or aperture of the other shell with its chelipeds and possibly also the walking legs
5. Grab crab in post-eviction/prechange stage – as for 1
6. Grasp shell in post-eviction/prechange stage – as for 2
7. Shell investigation in post-eviction/prechange stage – as for 4
8. Retreat in post-change stage – one crab moves away from the other
9. Cheliped extension in post-change stage – major and/or minor cheliped raised parallel to substrate and extended towards other crab
10. Grab crab in post-change stage – as for 1
11. Grasp shell in post-change stage – as for 2
12. Withdraw into shell in post-change stage – one crab withdraws deep into new shell so that only tips of chelipeds remain visible
13. Climb on shell in post-change stage – as for 3
14. Shell investigation in post-change stage – as for 4
15. Shell rocking in pre-eviction stage – attacker gently moves the defender's shell back and forth
16. Shell rapping in pre-eviction stage – the attacker hits its shell against that of the defender
17. Total fight time – time elapsed from defender withdrawing into shell until eviction
18. Cheliped flicking – the defender moves one or both chelipeds through a 30° arc in the aperture of its shell

between activities. These are between: investigation of the attacker's original shell and grasping of that shell; cheliped display to the evicted crab and grabbing the evicted crab; retreating away from the scene and withdrawing into the shell.

However, negative correlations also occur between activities in this stage, these being: cheliped display to the evicted crab and both grabbing the original shell and investigation of that shell. That is, the more the winner displays to the loser the less time the winner allocates to activities directed towards its original shell.

Fewer significant correlations occur between activities from different stages of the fight. In general, shell-directed activities tend to be positively correlated with each other and opponent-directed activities are also positively correlated. There are, however, few correlations between these two broad types of activity.

All the activities discussed above are performed by the attacker. The one activity recorded for the defender, cheliped flicking, is not significantly correlated with any activity performed by the attacker.

PRINCIPAL COMPONENTS ANALYSIS

Dowds (1983) used a principal components analysis to examine the 18 variables (Table 6.1) in order to reduce the number of variables to a few components or factors which it was hoped would make behavioural sense. In this way a better description of the data may be given and new hypotheses concerning causal relationships may be generated (Frey and Pimental 1978). For an example of this technique readers are referred to Huntingford (1976) and here we will not consider the various steps in the computation. Suffice it to say that the 18 variables (including total fight time) were reduced to six components that were considered to be worthy of note; these six accounted for 69% of the variance in the original data. Each of these components may be plotted as a histogram to reveal the 'loading' or extent of influence each of the original variables has on each new component (Fig. 6.17).

Component 1 accounts for 22.4% of the variance and is made up primarily by activities performed during the pre-eviction phase of the fight. Only loadings exceeding 0.6 in either direction are considered 'significant' (Frey and Pimental 1978). The significant loadings for this component are shell investigation, shell rocking, shell rapping and fight duration (up to eviction). Component 2 (12.1% of variance) is made up primarily by grasp shell and shell investigation in the post-change period of the fight. Component 3 (10% of variance) is made up primarily by grasp defender and shell investigation during the post-eviction/prechange period of the fight. This suggests that if the victor is to undertake a prolonged investigation of the evicted loser's shell prior to moving in then it may also have to engage in further combat with the loser. Component 4 (8.7% of variance) comprises grasping the defender's shell in the pre-eviction phase and climbing onto the attacker's old shell in

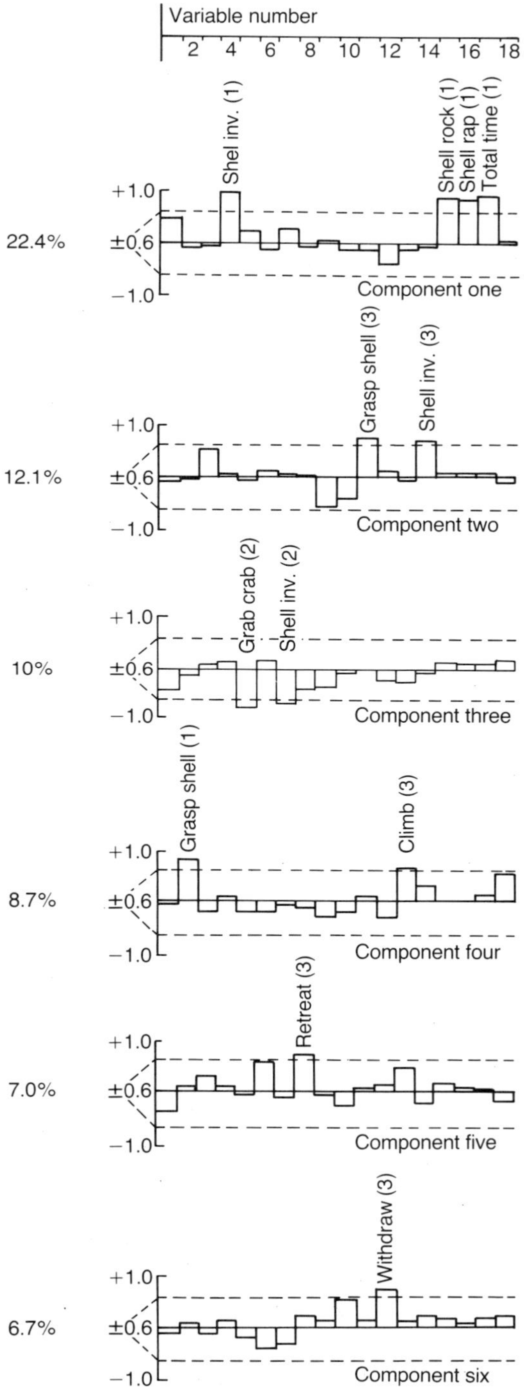

Variable number
2 4 6 8 10 12 14 16 18
22.4%
+1.0
±0.6
−1.0
Shell inv. (1)
Shell rock (1)
Shell rap (1)
Total time (1)
Component one
12.1%
Grasp shell (3)
Shell inv. (3)
Component two
10%
Grab crab (2)
Shell inv. (2)
Component three
8.7%
Grasp shell (1)
Climb (3)
Component four
7.0%
Retreat (3)
Component five
6.7%
Withdraw (3)
Component six

the post-change phase (both shell-oriented activities). Component 5 (7.0% of variance) contains only one variable, that of retreat by the attacker from the scene of the encounter. Finally, component 6 (6.7% of variance) includes only withdrawal into the new shell, an investigatory activity during which the new shell is tested.

The key point to emerge from this analysis is that it validates the intuitive splitting of the fight into three periods. The first three components, which together account for 44.5% of the variance, are each composed of activities from one period. Furthermore, each of these three components includes activities that may be described as shell assessment. This suggests that assessment of the opponent's shell is important in determinating the course of the fight up to the eviction of the defender (component 1). Further assessment of the defender's shell occurs after the eviction of the defender (component 3). After the victor changes shells there is a period of assessment of the victor's old shell (component 2). During this period there is also assessment of the newly acquired shell (component 6). The analysis thus indicates the importance of resource assessment in hermit crab fights. The analysis is less useful, however, in regard to helping with the interpretation of crab-directed activities. The role of shell rocking and shell rapping, both included in component 1, is not made clear by the principal components analysis. Neither is it clear whether the opponents assess each other's fighting ability or willingness to fight. Perhaps this is due to the omission of prefight displays from the analysis.

The multivariate analysis was therefore not completely successful in determining which factors influence shell fights. An experimental approach is required to determine which factors are assessed by the crabs and used in decision-making during shell fights. Such experiments, designed to elucidate the assessment abilities of crabs during shell fights, are discussed in the following chapter.

SUMMARY

Agonistic interactions, involving displays of chelipeds and walking legs, occur between crabs whilst disputing access to a variety of resources. Disputes over ownership of shells, however, are resolved by fights in which the attacker attempts to dislodge the defender from its shell. The fight may end with evacuation of the shell by the defender or by the attacker giving up. In the former case the attacker selects which shell to occupy. Principal components analysis (PCA) revealed the importance of activities involved in shell assessment by the attacker. PCA, however, did not reveal activities concerned with assessment of fighting ability.

Figure 6.17 The six components from the principal components analysis are shown as histograms to reveal the extent of influence of each of the original variables. Only variables with a 'loading' of greater than 0.6 in either direction are considered 'significant'. (From Dowds 1983.)

CHAPTER 7

Assessments during shell fights

A major advance in the study of aggressive interactions was made when the principles of game theory were applied to animal contests (Maynard Smith and Price 1973). Game theoretical analysis indicated possible resolutions to several points of confusion. Modern theory suggests that individuals will usually attempt to maximize personal gain in contests rather than acting for the good of the species, although exceptions may occur when close relatives compete for a resource (Hamilton 1964; Grafen 1979; Fry 1980). Natural selection acts in such a way that a population of animals comes to adopt an evolutionarily stable strategy (ESS): the strategy that, if employed by most members of the population, confers a higher fitness than any alternative strategy (within the limits of the organism's capacities or set of strategies defined in the game) (Maynard Smith and Price 1973).

A simple game that helps to clarify these concepts is the Hawk/Dove game (Maynard Smith 1976). In this game an animal may elect to play Hawk, which involves attacking the opponent and continuing to do so until injured or until the opponent retreats. Alternatively, the animal may elect to play Dove, which involves displaying for a predetermined period or, if attacked, retreating before getting injured. Should both contestants select the Hawk option one will be injured and incur the cost of a wound (W) and the other will gain the resource of a particular value (V). Should both select Dove, both will waste time in displaying at a particular cost (T) and one, arbitrarily, gains the resource. Should one select Hawk and the other Dove the Hawk will win. Thus Hawk might, at first, appear to be the best option, but whether selecting Hawk 100% of the time is best depends upon the relative values of the costs (W and T) and benefit (V) involved. An essential feature of this simple game is that it is assumed that all those playing Hawk are equal and all those playing Dove are equal. That is, the contests are essentially symmetrical and no assessment of the opponent takes place.

The average pay-offs for each of the four types of encounter are shown in the following matrix:

		Playing against	
		H	D
pay-off to:	H	$\frac{1}{2}(V-W)$	V
	D	0	$\frac{1}{2}V-T$

The ESS may be calculated as:

Play H with probability $(2T + V)/(2T + W)$
Play D with probability $1 - (2T + V)/(2T + W)$

Thus if we insert particular values for the cost of a wound (e.g. 20), for the cost of time wasting (e.g. 4) and the value of the resource (e.g. 10) the ESS is to play Hawk with a probability of 0.643 and play Dove on the other occasions. At these probabilities the pay-offs to Hawk and Dove are equal, both are favoured equally by selection and their probabilities are therefore evolutionarily stable.

For mathematical simplicity it is assumed that the costs and benefits remain constant from contest to contest, but this is a naïve assumption to make concerning the real world (Maynard Smith 1976; Parker 1974). Although the resources for which the animals are competing will vary in quality, if the animals have no means of assessing that variation selection will operate on the basis of the average benefit. Should different populations have different average benefits then the ESS will come to differ between those populations. For example, if one population has the value of the average resource as 16, and time and wound costs as above (i.e. 4 and 20 respectively), then the ESS for this population will be to play Hawk with a probability of 0.857. Thus as the value of the resource increases so it pays to play Hawk more frequently. Indeed, if the value of the resource is greater than the cost of a wound it pays to play Hawk 100% of the time. This is a pure strategy as opposed to the mixed strategies discussed above.

By definition the ESS is the optimal strategy because within the limits of an organism's capacities it cannot be beaten by an alternative strategy. It is obvious, however, that when the resource value varies between contests within a population, any animal that develops the ability to assess the variation could adjust the probability of playing Hawk and thus have a distinct advantage over others that operate on the average value. Thus the great advantage of assessment is made clear from this simple Hawk/Dove game (Parker 1974; Maynard Smith and Parker 1976). It would also benefit the animal if it could estimate the probability of receiving a wound (i.e. to assess the potential cost of the contest). It could do this if it could assess the fighting ability of its opponent. This fighting ability is usually described as the

Resource Holding Power (Parker 1974) or Resource Holding Potential (Maynard Smith and Parker 1976), both normally shortened to RHP. The important feature, is of course, the RHP of the opponent relative to that of the animal in question i.e. the relative holding power. Somewhat confusingly this term is also abbreviated to RHP (Riechert 1978), and sometimes treated as being the same as the above (Sigurjonsdittir and Parker 1981), although one term refers to the ability of a single animal whereas the other refers to the difference between two animals. Here the use of the term RHP is restricted to its original meaning and the term 'relative RHP' is used to indicate the difference between two animals.

Clearly, differences in relative RHP will lead to different potential costs in contests. Thus variation in relative RHP and in resource value should both influence the occurrence of specific acts within agonistic encounters, especially the occurrence of escalated fighting, i.e. adoption of the Hawk strategy. Alternatively, or in addition, these two variables might influence the duration of fights because increased duration is bound to lead to greater costs in terms of energy expended, accumulation of minor wounds and greater probability of receiving a major wound (Parker 1974). Such effects on the duration of fights are discussed in the following chapter. In the present chapter we consider whether or not real contests conform to the prediction that the occurrence of particular activities will vary according to the assessed costs and benefits (Maynard Smith and Parker 1976).

A number of studies have confirmed the existence of assessment strategies, particularly those assessing relative RHP. For example, male toads (*Bufo bufo*), assess their opponents prior to initiating a fight (Davies and Halliday 1977). Males with a low RHP, identified by the pitch of vocalizations, are more likely to be attacked than are males with a high RHP. Red deer stags also assess each other, partially on the basis of vocalizations and partially by vision (Clutton-Brock *et al.* 1979). Male dung flies competing for females also base decisions of whether or not to attack on the relative size of the opponent (Sigurjonsdottir and Parker 1981), indicating their ability to use distant, presumably visual, assessment. In other flies, however, this ability appears to be lacking, e.g. *Dryomyza anilis* (Otronen 1984) and *Harpobittacus nigriceps* (Thornhill 1984). In both of these species RHP assessment appears to take place after the fight starts. In spiders, *Frontinella pyramitela* lacks distant RHP assessment (Austad 1983) but in *Agelenopsis aperta* owners and intruders on webs apparently assess each other by vibration (Riechert 1978).

In this latter case, when closely matched individuals dispute ownership of a web, the contest starts with a prolonged series of displays with some escalation and de-escalation. The strategy appears to be similar to the 'Prober-retaliator' or 'Retaliator' strategies discussed by Maynard Smith and Price (1973). On the other hand, when the owner of the web is larger than the intruder, escalation by the owner is often immediate and the intruder usually retreats, i.e. similar to the 'Hawk' and 'Dove' strategies of Maynard Smith and Price (1973). The most complex interactions occur when the owner is smaller than the intruder and the best description of the strategies in simple game theory terms is that the intruder plays 'Hawk' whereas the resident plays 'Retaliator' (Riechert 1978).

The evidence that resource value influences the probability of attack is less abundant than that concerning relative RHP. Red deer stags tend to challenge harem holders rather than solitaries, i.e. situations in which a gain is possible (Clutton-Brock *et al.* 1979). The stomatopod *Gonodactylus viridis* is more likely to escalate an encounter if it is defending a cavity than if no cavity is held (Caldwell 1979; Caldwell and Dingle 1979). In the spider *A. aperta*, a web owner is more likely to escalate an encounter if the web is of good quality (Riechert 1978, 1984). The intruder, however, is unable to assess web quality directly and apparently monitors the defender's actions to gain some information. This feature of an intruder not being able to assess the resource value is quite common. It occurs in female iguanas contesting burrows (Rand and Rand 1976) and also in situations of mate competition in arthropods in which one male has information concerning the amount of sperm transferred but the other does not, e.g. the spider *F. pyramitela* (Austad 1983) and the dung fly *Scatophaga stercoraria* (Sigurjonsdottir and Parker 1981). In each case this gives the 'owner' of the resource more information and thus greater opportunity to modify its behaviour in contests. In other cases of intruders upon mating pairs resource quality has little or no effect, as in the fly *D. anilis* (Otronen 1984), or both contestants have good information, as in the red-spotted newt (Verrell 1986) and the fresh water amphipod, *Gammarus pulex* (Dick and Elwood 1990). However, although these studies demonstrate assessment abilities these assessments do not always lead to changes in the occurrence of specific activities.

ASSESSMENTS DURING SHELL FIGHTS: EFFECTS OF RESOURCE VALUE

We have seen that certain parameters of shell quality influence the likelihood of crabs engaging in and winning agonistic encounters involving displays other than shell fights (Chapter 6). Shell quality also has marked effects on the initiation and outcome of shell fights. The addition of weights onto the shells of *Clibanarius vittatus* makes crabs more likely to initiate and win shell fights (Hazlett 1970a). Furthermore, experimentally damaging a crab's shell also increases shell fighting (Hazlett 1970a). In a study of *P. bernhardus* in which crabs were allowed to engage in fights it was found that crabs initiated fights in which either a larger or a smaller shell might be gained by the attacker (Hazlett 1970a). Using covariance analysis it was found that crabs fighting for a larger shell appeared to be attempting to gain a greater volume rather than a greater weight. Conversely, those fighting for smaller shells appeared to be fighting for a shell of less weight rather than a shell of smaller volume. Thus crabs appear to be fighting for shells that are light in weight but of sufficient volume for their body size.

These studies indicate features of shells that make shell fights more likely but give little insight into how the course of the fight is influenced by the resource value. An experiment with this aim was conducted on *P. bernhardus* in which large and small crabs were paired (Dowds and Elwood 1983). The size difference was standardized but the species of shell varied among four experimental groups. In the first group the large crab was in an unpreferred *Gibbula* shell (Chapter 3) whereas the small crab was in a preferred *Littorina obtusata* shell (GL group). The larger crab could make a gain in shell preference if it could evict the smaller crab. In the second group both crabs were in *Gibbula* shells (GG group) and in the third group both were in *L. obtusata* shells (LL group). In these two latter groups no gain or loss could be made by the large crab evicting the smaller. In the final group the large crab was in *L. obtusata* whereas the small crab was in a *Gibbula* shell (LG group) and in this case the large crab would lose in shell quality if it exchanged shells. Forty pairs of crabs were observed in each group and all shells used were of the preferred size for the larger crab (Chapter 3) (i.e. too large for the smaller crab). In this manner we could determine how the qualities of both the attacker's and defender's shells influence decisions during shell fights.

Of the 125 fights observed in this experiment 93.3% were initiated by the larger crab. The role of the relative size of the crabs is considered in the following section, so suffice it to say here that the few fights initiated by the smaller crab were considered aberrant and are disregarded in the present analysis.

The main results of the experiment are shown in Fig. 7.1, with the key point being the variability in the number of evictions among the groups. Shell quality obviously had a marked effect on decisions in these fights and we now examine these decisions, first for the attacker and then for the defender.

The attacker – to fight or not

If the larger crab could assess the quality of the opponent's shell then it would pay to attack if by that action it might acquire a higher quality shell. By examining the number of fights initiated by the larger crab in each group it is possible to determine whether the smaller crab's shell is assessed (Table 7.1), specifically by comparing the GL group with the GG group and the LG group with the LL group. These comparisons show no difference in the pro-

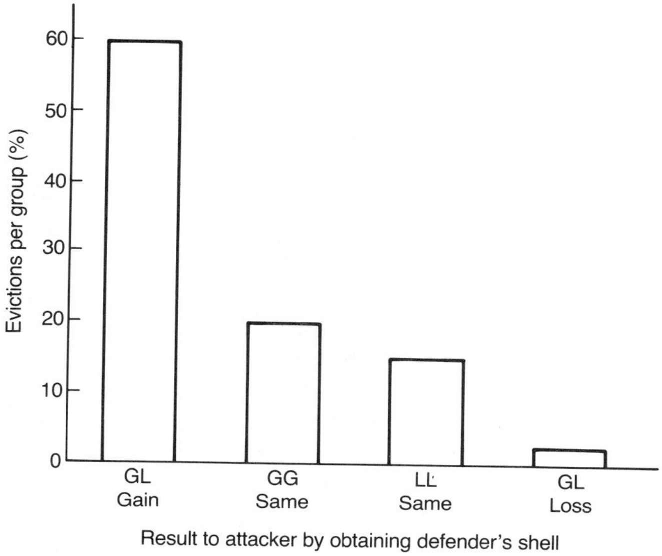

Figure 7.1 The percentage of evictions is considerably greater when the attacker may gain by obtaining the defender's shell.

bability of attack which suggests that the attacker is incapable of assessing the opponent's shell from a distance. In this respect the data are in agreement with Reese (1963) who also reported a lack of ability to discriminate species of shell from a distance. In Chapter 5, however, experiments were described that indicated that crabs could gain information about shell quality from a distance, specifically that differences between the species of shell used in the current experiments could be assessed. Perhaps the presence of another crab in the aggressive situation makes it more difficult to assess the other shell.

Table 7.1. Influence of the smaller crab's shell on the initiation of shell fights by the larger crab

Group	*Fight*	*No fight*	
GL	32	4	N.S.*
GG	32	4	
LG	28	12	N.S.*
LL	33	6	

*Not significant.

This, however, is surprising as a distant but accurate assessment of the resource could dispense with unnecessary shell fights.

In the light of previous experiments (Hazlett 1970a, c) it might be expected that the quality of the attacker's shell should influence the decision of whether or not to initiate a fight. The data for the appropriate comparisons of experimental groups are shown in Table 7.2. No significant difference occurs between the GL and LL groups, although, when the LG and GG groups are compared it is seen that crabs in poor-quality shells are more likely to initiate fights ($P = 0.03$). A much more marked effect was noted by Neil (1985) who observed pairs of crabs in which one of each pair was in a shell with a partially occluded aperture. This prevented the crab from withdrawing into the shell and thus made it far from satisfactory. Only 5 out of 80 crabs in normal shells initiated shell fights compared to 57 out of 80 in modified shells ($P < 0.001$) indicating that shell quality can have major effects on the decision of whether or not to attack. The lack of such an effect was reported for *P. hirsuitiusculus* (Vance 1972b) but the shells used in that study may not have been sufficiently different from the preferred size to influence the behaviour markedly.

The attacker: to escalate the fight or not

Once a fight has started the attacker has a choice of whether or not to commence shell rapping. In the study of Dowds and Elwood (1983) the attacker engaged in shell rapping in 36% of fights (45/125) and eviction of the defender occurred in 86% of fights in which rapping occurred. It seems reasonable to assume that this relatively vigorous activity is an escalation of the fight to a more intense level than that seen at the start of the fight. Neil (1985) referred to level 1 and level 2 fights, the latter being fights that included shell rapping. Other components of escalation involve shell rocking and grabbing at the defender, but here the term escalation is restricted to shell rapping. We use the term 'escalation' in the same way as other authors

Table 7.2. Influence of the larger crab's shell on its initiation of shell fights

Group	*Fight*	*No fight*	
GL	32	4	N.S.*
LL	33	6	
GG	32	4	$P = 0.03$
LG	28	12	

*Not significant.

to suggest a more vigorous fighting activity than is seen in displays (e.g. Clutton-Brock *et al.* 1979).

Although the quality of the defender's shell does not seem to influence the probability of a fight occurring (see above) it does influence the probability that the attacker will escalate fights. The attacker was considerably more likely to escalate if the defender was in a good quality *L. obtusata* shell than if it was in a lower quality *Gibbula* shell (Table 7.3). Furthermore, escalation was more likely to occur if the attacker was in a poor quality shell (Table 7.4). These data demonstrate that resource assessment plays a major part in the decision by attackers of whether or not to escalate the fight. Clearly, information concerning the quality of the defender's shell must be gathered during the early stages of a shell fight, between the time of grabbing the shell and the decision to escalate or cease fighting.

Elwood and Glass (1981) hypothesized that if the attacker could gain by exchanging shells then escalation would be more likely to occur. Abrams (1982a, b), however, hypothesized that attackers would be more willing to escalate only if their own shell quality is low. These alternative hypotheses are easily tested with the data of Dowds and Elwood (1983). The first predicts probability of escalation occurring in fights in the following order:

$$\text{GL} > \text{GG} = \text{LL} > \text{LG}$$

Table 7.3. Influence of defender's shell on the escalation of shell fights

Group	*Escalate*	*Non-escalate*	
GL	24	8	$P < 0.01$
GG	10	22	
LL	9	24	$P = 0.03$
LG	2	26	

Table 7.4. Influence of the attacker's shell on the escalation of shell fights

Group	*Escalate*	*Non-escalate*	
GL	24	8	$P < 0.001$
LL	9	24	
GG	10	22	$P = 0.02$
LG	2	26	

The second predicts a different order:

$$GL = GG > LL = LG.$$

The data for the percentage of fights that were escalated for the four groups were:

	GL	GG	LL	LG
Escalate (%)	75.0	31.0	27.3	7.1

These data clearly uphold the hypothesis of Elwood and Glass (1981) and indicate that during the early stages of shell fights the attacker obtains information about the quality of the defender's shell, combines it with information about the quality of its own shell and thereby assesses the gain which could be made.

The attacker: to evict the defender or not

We will examine in the following chapter the respective roles of attacker and defender in determining whether or not the defender is evicted. In the present section we will simply describe the factors concerning shell quality that influence the probability of eviction. We have already stated that the majority of escalated fights result in eviction (at least in the study of Dowds and Elwood (1983)) and thus the pattern for eviction follows that of escalation (Fig. 7.1). Significantly more evictions occur if the attacker is in a poor-quality shell (GL vs LL, $P < 0.001$; LG vs GG, $P = 0.02$) and also if the defender is in a good-quality shell (GL vs GG, $P < 0.001$; LG vs LL, $P = 0.07$). Eviction of the smaller crab is thus more likely if the larger crab gains in shell quality.

The attacker: to change shells or not

In the study of Dowds and Elwood (1983) all successful attackers eventually moved to the vacated defender's shell, but only after they had examined it. In a study in which some defenders had been in particularly poor quality shells (partially occluded apertures) these defenders moved out of their shells prior to them being perceived as poor quality by the attackers. After eviction, however, the attackers did not move into those shells (Neil 1985), so clearly further information about shell quality is gathered after the defender is evicted but prior to the victor moving in (see also Chapter 6).

After moving into the new shells the crabs usually assess the fit by withdrawing deeply into them. They also investigate their old shell (Dowds and Elwood 1983) and some crabs then move back to their original shells. We may predict that the probability of remaining in a newly acquired shell should reflect the assessment of gain from the new shell relative to the old. Thus we predict the following order for permanency of shell change:

$$GL > GG = LL > LG.$$

The percentage of each group remaining in the new shell was:

	GL	GG	LL	LG
Number that changed to new shell:	24	8	6	1
Percentage of above that remained in the new shell:	100%	62.5%	33%	0%

Considering the small numbers involved in some groups this conforms to the predicted result. It demonstrates that the assessments of shell quality in the early stage of fights, although good, are not perfect. The single crab that made a loss in moving from the preferred species to the unpreferred species quickly moved back whereas all twenty-four crabs moving from *Gibbula* to *L. obtusata* shells remained in those shells.

At this point in a shell fight the victor releases hold of the unwanted shell and allows the defender to enter it. It might appear from these analyses that the defender is passive in these interactions. That, however, is not the case.

The defender: general

We have seen that the attacker acts apparently to maximize its gain during shell fights and theoretically the defender will attempt to do the same. Indeed, it has been suggested that the defender has a more important role than the attacker in determining the duration (Hazlett 1978) and consequent result of shell fights (Hazlett 1981). It has been proposed that the defender gains information about both shells and bases its decision, in part, upon their relative qualities, i.e. on whether or not the defender might gain from an exchange. Hazlett (1978) suggested that these interactions are described more accurately as 'negotiations' rather than 'fights'. The data of Dowds and Elwood (1983) allow for this hypothesis, concerning the outcome of interactions (Hazlett 1981a), to be tested. Other data are required to test the hypothesis that durations of fights are influenced primarily by the defender and these data are discussed in the following chapter.

The defender: to fight or not

The optimal strategy for a small crab in a good shell on encountering a large crab might be to retreat rapidly and thus avoid a fight. If however, the smaller crab is in a poor shell, the better strategy may be to withdraw into the shell and allow a shell fight to start. This strategy could be particularly advantageous if the defender could assess the quality of the potential attacker's shell from a distance and thus take this into account in deciding whether or not to allow a fight.

Dowds and Elwood (1983) examined these possibilities but found no evidence that small crabs in poor shells were more willing to engage in shell fights than those in better shells. Concerning the quality of the attacker's shell, small crabs are more likely to be attacked if the attacker is in a poor shell (which is poor also for the defender) rather than a good shell. Thus there is no evidence to suggest that small crabs are more willing to enter fights when it might pay them to do so. It thus seems that they too are unable to make the assessments that would enable the optimum strategy to be employed. In the absence of information concerning the larger crab's shell the optimum strategy would seem to be to attempt to avoid fights, particularly if in a good shell, but this is not always possible, either in the confines of experimental situations or in natural situations. It might be predicted that when the small crab is carrying a shell far too large and heavy its mobility would be impaired. There is, however, no unequivocal evidence to support this prediction.

The defender: to cheliped flick or not

The major defensive strategy of a hermit crab is to remain tightly withdrawn into its shell, but it is capable of one distinctive defensive action, cheliped flicking. Hazlett (1970c) described and investigated the use of this action and concluded that it increases the probability that an attacker will release its grip on the shell of the defender. The flicks are produced when the defender is withdrawn into the shell and appear to be initiated by tactile stimuli (Hazlett 1970c). Dowds and Elwood (1983) considered whether the quality of one or of both shells influenced the occurrence of this activity.

It is predicted that defenders will have knowledge of their own shell quality and defend good quality shells more than poor quality shells. In the study by Dowds and Elwood (1983), 38% of small crabs in *L. obtusata* shells (good quality) showed cheliped flicking during fights compared with 10% of crabs in *Gibbula* shells (poor quality) ($\chi^2 = 12.06$; $P < 0.001$). Thus the above prediction appears to be upheld. It might simply be, however, that flicks are more common in escalated fights and escalated fights are more common when the defender is in a good shell. If only these escalated fights are examined we find that 57% of defenders in good shells and 25% of those in poor shells flick their chelipeds ($P = 0.04$, Fisher test). A similar trend is found for unescalated fights but is not significant. Thus we can confirm that good shells are defended more than poor shells.

It would be to the defender's advantage if it could adjust the probability of cheliped flicking according to the quality of the attacker's shell. That is, it should defend less if the attacker is in a good shell if there is some prospect of getting that shell from the attacker. There is a non-significant trend in the predicted direction but this appears to be due to the higher probability of escalation if the attacker is in a poor shell (Dowds and Elwood 1983). When escalated fights are examined separately no trend exists and we conclude that defenders can only assess their own shell, not the attacker's shell. Cheliped flicking is more likely to occur in escalated fights than in unescalated fights (48% vs 11%, $\chi^2 = 19.9$, $P < 0.001$).

The defender: to evacuate the shell or not

At some time during a shell fight the defender might release its abdominal hold on the shell and come up in the aperture. Often the attacker appears to pull at the appendages for some time prior to this 'coming up' and then the defender is usually pulled out of the shell (Chapter 6). Factors influencing the decision of *when* to evacuate the shell are considered in the following chapter. Here we consider the effects of the relative shell qualities on the probability of evacuation. Hazlett (1981a) suggested that an evacuation would be more likely if the defender gains in terms of shell quality by an exchange. Thus in the experiment of Dowds and Elwood (1983) the following order for evacuation is predicted:

$$\text{LG} > \text{GG} = \text{LL} > \text{GL}.$$

Alternatively Elwood and Glass (1981) have suggested that a shell change is more probable when the attacker may gain from exchange. Thus the opposite order to the above is predicted:

$$\text{GL} > \text{GG} = \text{LL} > \text{LG}.$$

The data show a clear fit to the hypothesis of Elwood and Glass (1981):

	GL	GG	LL	LG
Number of fights	32	32	33	28
Fights ending in eviction (%)	75%	25%	18%	4%

Thus it appears that the attacker is capable of making the more accurate assessments during shell fights, at least concerning shell qualities. The attacker gains information about both shells, the defender only about its own shell. This asymmetry in information gives the attacker a distinct advantage. The defender would obviously be able to adjust its behaviour more accurately if it had information about the attacker's shell but it seems unable to obtain that information. There is the possibility that this is achieved in other species, however (Hazlett 1981a, 1987).

ASSESSMENTS DURING SHELL FIGHTS: EFFECTS OF RELATIVE RESOURCE HOLDING POWER

In the study of resource value (Dowds and Elwood 1983) 93.3% of the fights were initiated by the larger crab. It is thus obvious that some assessment of the relative sizes, and hence assumed fighting abilities, occurs prior to the initiation of a fight. Indeed, the initial displays may be a means of determining which is the larger of the two crabs and hence which roles (attacker or defender) they will assume in a fight. Of the nine fights initiated by smaller crabs in that study all smaller attackers retreated after a period of aperture investigation, that is after a time when the defender's chelipeds were visible. Thus assessment of the size of the defender probably continues during the fight. It is also probable that the defender continues to assess the fighting abilities of the attacker, perhaps by the power of the shell rapping and the pulling at its appendages. In this section we examine the evidence that relative RHP influences decisions made during shell fights. However, we should first clarify some of the difficulties in performing such studies using hermit crabs.

In many animal fights the single resource in dispute is of the same value to each contestant (e.g. newt struggles for mates, Verrell 1986). In other situations the value of the resource varies for each of the contestants in some relationship to their size. Examples of this concern struggles for mates in toads (Davies and Halliday 1977) and contests for burrows in stomatopods (Dingle 1983). In these latter examples the value of the resource is different for the two contestants. The situation in hermit crab contests, however, is even more complex because there are two resources rather than one.

We have already seen how the value of each shell is assessed by the attacker and how this influences the attacker's decisions. The value of the defender's shell also influences the defender's decisions. The difficulties arise because, as the relative sizes of the contestants vary from contest to contest, so the value of both shells for each of the participants will change. Thus effects due to relative RHP and resource value may not be fully separable in any one experiment.

There have been two approaches that attempt to circumvent these problems. First, the quality of the shell of one of the participants in a fight can be reduced to such a low level that obtaining almost any shell from a defender will represent a considerable gain. Thus defenders, regardless of their size, can be given preferred shells for them and yet these shells will be valued highly by attackers regardless of their size (within reasonable limits). This has been achieved by partially occluding the aperture of the shell of one of the participants making it difficult for the crab to withdraw (Neil 1985) or by removing the shell completely so that a naked crab is fighting for the only shell in the contest (Elwood and Glass 1981; Dowds and Elwood 1985).

Second, it might be argued that variations in resource value might influence the course of the fight up to the eviction of the defender. After eviction, however, the size of the naked 'defender' relative to that of the victor might influence the behaviour of the crabs. Recall that after eviction the victor selects which shell to take and this process might be influenced by the relative sizes of the crabs.

To fight or not to fight

Neil (1985) paired crabs so that one of the pair was in a shell with a partially occluded aperture whereas the other crab was in a shell of its preferred size. Crabs in the modified (poor) shells were much more likely to initiate a fight if they were larger than the other crab than if they were smaller. Similar effects were noted when naked crabs were paired with housed crabs (Dowds and Elwood 1985) (Fig. 7.2).

The size of the chelipeds appears to be assessed during the prefight displays (Chapter 6) and this assessment has been investigated with specific regard to shell fights (Neil 1985). Normal crabs, crabs with regenerating and hence small major chelipeds and crabs that had their major chelipeds removed were compared with respect to shell fighting ability. These crabs were induced to occupy shells with partially occluded apertures (suboptimal) and paired with crabs in normal shells. Crabs with missing major chelipeds were less likely to engage in contests than were intact crabs and those with intermediate size chelipeds (regenerating) showed intermediate behaviour. Furthermore, smaller crabs were more likely to establish dominance over larger ones if the

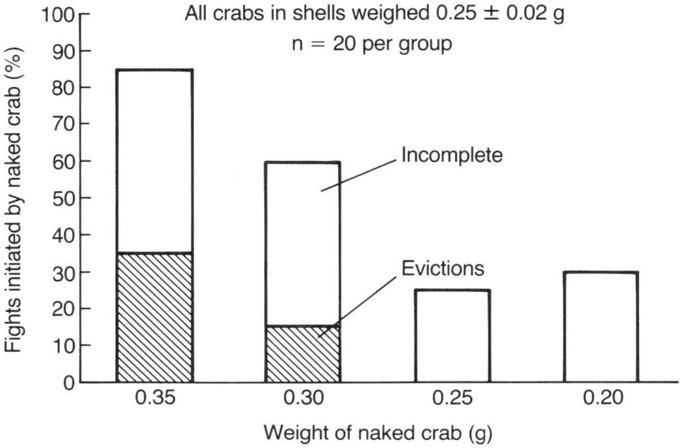

Figure 7.2 More fights were initiated by the naked crab when its was larger than the housed crab and evictions occurred only when the naked crab was larger.

latter had small or missing chelipeds. The data suggested that the crab with the small or missing cheliped had a decreased motivation to fight whereas a crab perceiving such an opponent had an increased motivation to fight. That is, the major cheliped of the opponent appeared to be assessed relative to the crab's own cheliped.

The position of the crabs may also have temporary influences on the assessment of relative size. Crabs that are positioned on a higher level may have an advantage, and certain displays increase the apparent height of the crab (Chapter 6). Crabs may also attack from the side or rear and thus not allow their opponent time for RHP assessment.

Decisions during the fight

The relative RHP of the crabs also influences decisions made in the course of the fight. Thus defenders that lack their major cheliped are more likely to be evicted than are crabs with normal chelipeds (95% vs 25%, $P = 0.0002$) indicating that the cheliped is important in defence (Neil 1985). Cheliped flicking by defenders occurs predominantly in escalated fights and may be a response to certain tactile stimuli as the attacker attempts to grasp the defender. The defender, however, does not respond to just tactile stimuli, rather it appears to remember its assessment of the opponent which was made before it withdrew into the shell (Hazlett 1970c). Crabs that withdraw in response to the approach of a large crab are less likely to flick their chelipeds in response to being touched on the chelipeds (by the experimenter) than are crabs which withdraw in response to an opponent of equal size. This demonstrates that information gained at one stage in a fight influences later strategic decisions.

Relative size also has a marked effect on the probability that a fight will be escalated. Naked attackers that are larger than the defenders are more likely to escalate the contest, by hitting their naked abdomens against the defender's shell (equivalent to shell rapping), than are those that are the same size or smaller (98% vs 55%, $P = 0.0035$) (Dowds and Elwood 1985). The attacker may remember its prefight assessment of the defender but it also has the opportunity to gain further information from visual and tactile cues during the initial stages of the shell fight. Attackers are more likely to escalate against defenders with small (regenerating) or missing major chelipeds than against normal crabs (Neil 1985).

Fights are more likely to result in an eviction if naked attackers are larger than the defender as opposed to being smaller or the same size (36% vs 0%, $P = 0.024$; Dowds and Elwood 1985) and similar size effects occur with respect to attackers in markedly suboptimal shells (Neil 1985).

It thus appears that crabs assess their opponent and their chances of winning an encounter, and use this information as a basis for decisions during the fight. Relative size can also influence decisions after the defender has been evicted. At this time the victor attempts to assess the new shell but if the defeated opponent is relatively large then one component of this resource assessment, withdrawing into the new shell, is less likely to occur than if the defeated opponent is relatively small (Dowds and Elwood 1985). In omitting this activity the victor may leave the vicinity of the fight sooner and so avoid the possibility of a further interaction.

COMBINED EFFECTS OF RHP AND RESOURCE VALUE ASYMMETRIES

The various factors that influence a number of visible decision points during hermit crab fights are summarized in Fig. 7.3. The data indicate the following conclusions. Two features are assessed in arriving at the decision of whether or not to attack, these being the value of the crab's own shell and the relative size of the opponent as judged by the size of the major cheliped. Whether or

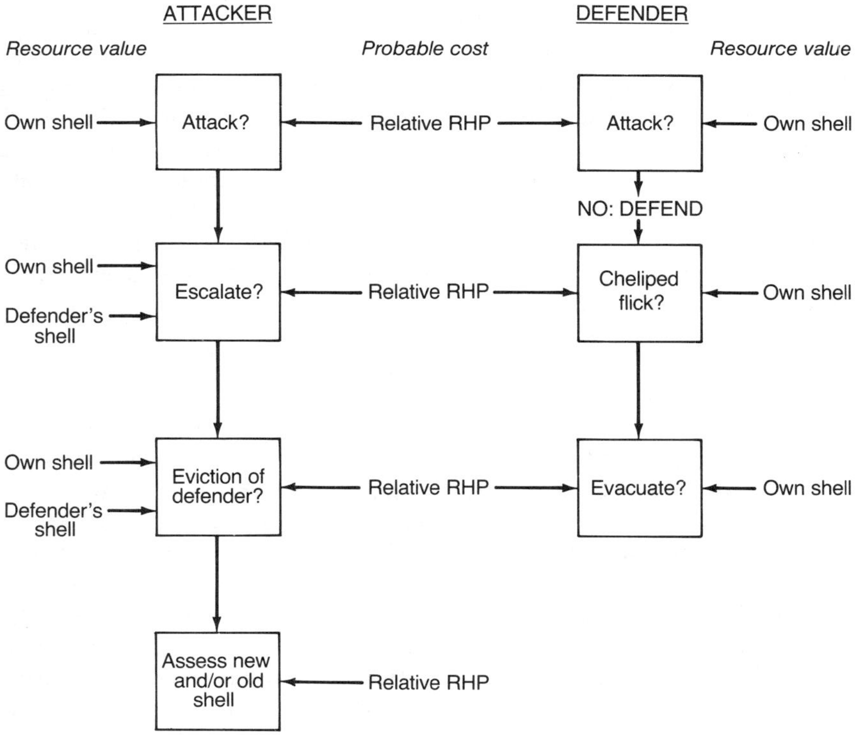

Figure 7.3 Sources of information used in decisions during fights.

not to escalate is influenced by the value of the attacker's shell compared to that of the defender. Relative RHP also influences this decision. Furthermore, relative RHP and the quality of the defender's shell influence the probability that the defender will employ cheliped flicking. Defenders are more likely to be evicted if the attacker has relatively high RHP and if the attacker may assess a potential gain from the resource. The potential gain to the attacker from obtaining the defender's shell also influences the probability of remaining in that shell rather than moving back to the original shell. Finally, relative RHP influences the probability of further assessing the new shell after shell change.

One feature of these contests that becomes clear from this analysis is that not all information is available at the start of the fight. This is particularly true of the resource value. At the initiation of a contest each crab acts as if it only has information about its own shell quality. In the early stages of the fight proper the attacker, but not the defender, gains information about the other shell and thus can assess more accurately the gain that it may make. This gives the attacker a distinct advantage over the defender in making sound decisions.

These experiments have identified a variety of factors that influence particular visible decisions. However, these factors alone do not appear to *determine* the outcome of the shell fight. Rather, the outcome appears to be determined by the duration for which each contestant is prepared to fight. At any point in the interaction the attacker may cease attacking, leaving the defender as the apparent winner. Alternatively the defender may evacuate its shell leaving the attacker as the apparent winner and able to make its choice of the two shells. The victor is thus the crab that persists the longest in fights and this persistence and the timing of decisions in general is discussed in the following chapter.

SUMMARY

Experiments in which shell quality of attackers and defenders was varied so that each might gain, lose or have no change in resource value by shell exchange are reviewed. These demonstrate the sources of information to attackers and defenders and when that information is gained. Attackers gather information about the defender's shell but the defender is unable to obtain information about the attacker's shell. The potential gain in resource value to the attacker is a major determinant of decisions during a fight. The defender has less information and decisions on whether to cheliped flick (a defensive activity) are based on the quality of the shell occupied. The relative size of the contestants also influences decisions by both contestants. These sources of information combine to determine the nature and outcome of the fight.

CHAPTER 8

Assessment strategies and the duration of fights

During a shell fight either the attacker gives up and leaves or the defender releases its grip on the shell and, after signalling to the attacker by tapping its chelipeds and legs against the attacker's appendages, allows itself to be drawn out of the shell. The attacker then makes its choice of shells. A key feature in determining which of the combatants wins the fight is the time for which each is prepared to persist. Fights determined by persistence have been termed 'wars of attrition' (e.g. Maynard Smith and Parker 1976) and in this chapter some theoretical and empirical examples are examined and compared to our observations on hermit crabs. In the following chapter a motivational model is developed that is compatible with the functional analysis.

The symmetrical war of attrition (Maynard Smith and Price 1973, Maynard Smith 1974, 1976) assumes that the two contestants have equal abilities and that no assessment takes place. The contest is settled purely on the basis of which is prepared to invest more in the contest in terms of time. Thus, if one contestant competes for time m_1 and the other for time m_2, where $m_1 > m_2$, then the pay-off to the victor is $v - m_2$ and to the loser is $- m_2$. Note that the victor has to fight only up to the point where the loser quits. The problem is to find an ESS for the persistence time.

There is no pure ESS in this model. Suppose a population adopts a strategy of playing m; then any mutant that played slightly more would win. There will be selection for increasing persistence times. The expected benefit from a fight is $v/2$ (as only one animal gains the resource). If m exceeds $v/2$ the animals would, on average, make a loss from each encounter and playing $m =$ zero would result in a higher fitness. It might seem that playing $m = v/2$ is the ESS but a variable strategy with a distribution of allocated fight times, following a negatively exponential distribution, turns out to be the only stable state (Maynard Smith 1974). Because actual fight times are determined by the shorter bid the distribution of actual fight times is even more biased towards a predominance of short durations. The distribution of bids is such that the mean cost of the contests will be $v/2$, that is the expected pay-off to each contestant is 0 (Parker 1984). Some studies have shown a remarkable congruity with this predicted distribution, e.g. those on bowl and doily spiders (Austad 1983), cichlids (Neil 1983) and hermit crabs (Elwood and Neil 1986b) (Fig. 8.1). In these cases, however, there is good reason to believe that the contests are not symmetrical and that assessment takes place so that the data are not compatible with the symmetrical war of attrition model. Before continuing with more realistic models two points should be made about the symmetrical model. First, the bid times are selected and fixed before the fight; there is no assessment and consequent change of bid during the fight. Second, cues that might enable assessment of bid duration should be eliminated since clues to giving up time could be exploited by the opponent (Maynard Smith 1974).

ASYMMETRIES IN THE WAR OF ATTRITION

No unambiguous case of a symmetrical war of attrition has been described either from field or laboratory studies. This reflects the widespread occurrence of asymmetries that are assessed by the contestants and used in the determination of bid durations and fight outcome (Maynard Smith and Parker 1976). These asymmetries fall under three headings.

Resource value asymmetries

The average bid in the symmetrical model is expected to result in an expenditure equivalent to half the resource value. Thus if the resource value varies among fights, an animal able to assess this variation should adjust its bid

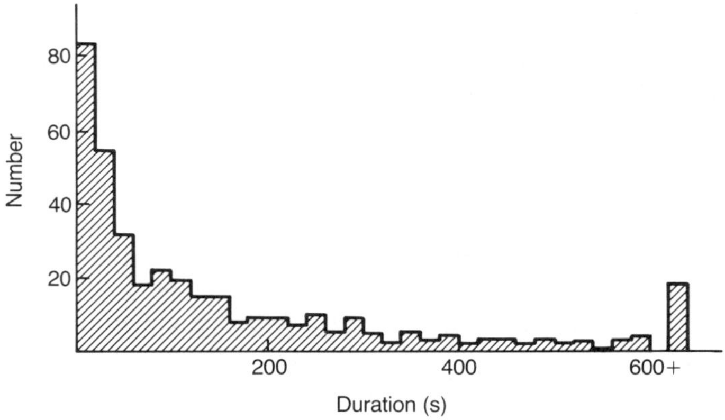

Figure 8.1 Distribution of durations of all fights. The data conform to a negative exponential (From Elwood and Neil 1986b).

and there is good evidence that animals can make this type of assessment. For example, in fights between male newts for females, larger females produce longer fights (Verrell 1986). Larger females have more eggs and hence are a more valuable resource (Fig. 8.2). Resource value also appears to influence fight duration in red deer, as fights between males for access to females increase in duration in mid rut, i.e. when fertilization is most likely (Clutton-Brock *et al.* 1979). In these examples there is no asymmetry in resource value as it has the same value for each contestant. In contests over food, however, it is easy to see that a hungry animal will value the food more than one that is satiated and these types of asymmetry may be used to settle disputes (Parker 1974; Hazlett *et al.* 1975).

RHP asymmetries

If information about relative RHP is perfect, then all other things being equal, the individual with the lower RHP should retreat prior to combat and thus avoid wasting time and risking injury (Parker 1974). Where RHP assessment is not perfect, however, a contest may be required to settle the dispute, and this is particularly likely where the opponents are of similar RHP (Parker 1974). These predictions have been upheld in numerous studies, for example in dung flies (Sigurjonsdottir and Parker 1981), scorpionflies (Thornhill 1984), the fly *Dryomyza anilis* (Otronen 1984) and two different spiders

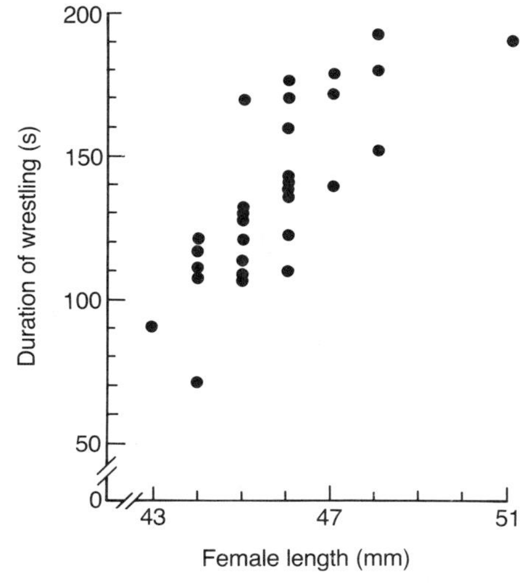

Figure 8.2 The relation between duration of wrestling (s) and female snout-vent length (mm) in red-spotted vents when owners and intruders are matched for length (From Verrell 1986.)

(Riechert 1978, Austad 1983). The examples of scorpionflies (Fig. 8.3) and bowl and doily spiders (Fig. 8.4) are particularly striking. They both indicate that distant RHP assessment is not accurate and that even when the size difference is great a short interaction is required. The key finding of both studies is that size, and hence RHP asymmetry, markedly influences fight duration.

RHP asymmetry may influence fight duration because the individual with the higher RHP can allocate more time or energy to the fight (Rand and Rand 1976; Riechert 1978) or has a higher probability of inflicting (rather than receiving) injury (Austad 1983). Austad demonstrated in the bowl and doily spider that the risk of injury rose linearly with fight duration, that the larger the size difference the more likely an injury was and that these injuries were more likely to be inflicted on the smaller opponent. This enabled Austad (1983) to calculate the cost of fights in terms of loss in reproductive success (Fig. 8.5), a considerable advance over most other studies of aggression. It should be noted, however, that not all studies of contests have found either duration or outcome to be correlated with size asymmetry (e.g. Harvey and Corbet 1986).

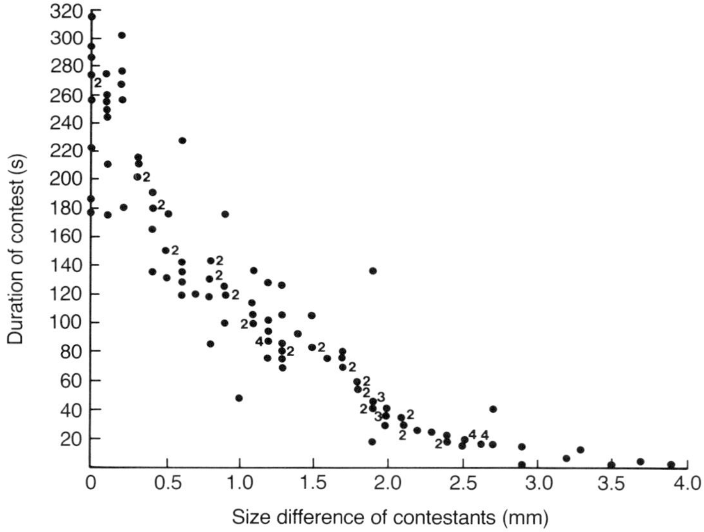

Figure 8.3 The relationship between the duration of fights and the size difference of contestants (forewing length in mm) in 125 natural fights in the scorpionfly *H. nigriceps*. The relationship is curvilinear and strong. The numbers to the right of dots indicate more than one observation. (From Thornhill 1984.)

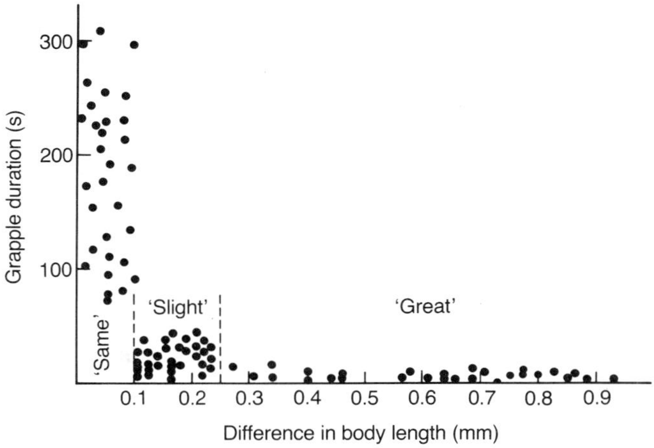

Figure 8.4 Grapple duration as a function of size asymmetry in the bowl and doily spider. These data are from 104 staged fights between males simultaneously introduced into the webs of virgin females. One outlier (0.05, 604) is not shown. The dotted lines indicate boundaries of the size-difference categories as used in subsequent experiments. These size-difference categories were determined by visual inspection of the data. (From Austad 1983.)

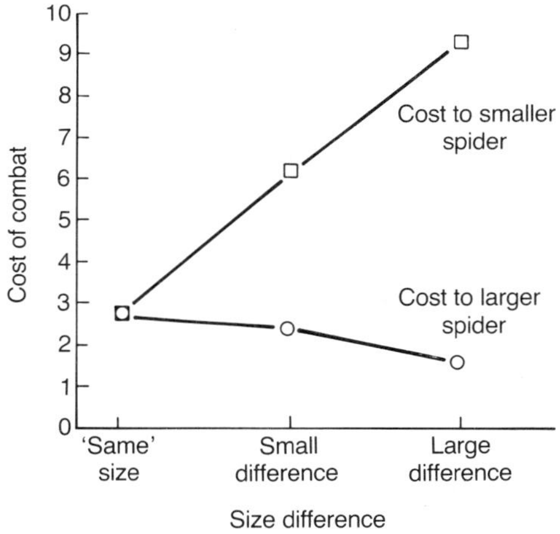

Figure 8.5 The cost to the 'smaller' spider in fights increases markedly as the size discrepancy increases and that for the larger spider decreases. (Data taken from Austad 1983.)

Owner–intruder asymmetries

It is a common situation in contests that one animal is in possession of the resource prior to the approach of an opponent. This type of asymmetry is called an owner–intruder asymmetry and it appears to be important in settling disputes (Maynard Smith 1974). It is common, though not universal, that owners have a greater probability of winning, e.g. in spiders (Riechert 1978), scorpionflies (Thornhill 1984), dungflies (Sigurjonsdottir and Parker 1981), speckled wood butterflies (Davies 1978), dragonfly larvae (Harvey and Corbet 1986), newts (Verrell 1986), iguanas (Rand and Rand 1976) and fulmars (Enquist *et al.* 1985). This feature of the owner winning might be explained in a number of ways and various hypotheses have been put forward. These hypotheses are discussed below, although more thorough analyses may be found in Hammerstein (1981), Leimar and Enquist (1984) and Enquist and Leimar (1987).

Owners are better quality individuals

If the resource typically lasts for sufficient time for it to be contested several times and the winner is determined on the basis of relative RHP then owners of the resource will be likely to be of high quality. That is, if original ownership is random with respect to RHP those owners with low RHP will be ousted by intruders with high RHP (Leimar and Enquist 1984). The key point of this scenario is that the owner–intruder asymmetry is irrelevant in any one contest, is not assessed and is not used to determine the winner.

Ownership determines victory by convention

Conventions may be used to determine the outcome of fights; that is, a convention may be an ESS (Maynard Smith 1974). This requires a conditional strategy of the type: 'if owner, fight for period M, where $M > v$ (where v = resource value), but if intruder bid 0'. This is superior to an alternative unconditional strategy of 'always fight for period m. If $M > m$ the pay-off for the two strategies on meeting with each other is the same ($1/2\,v - 1/2\,m$). However the pay-off when two conditional strategists meet is $1/2\ v$, which is higher than when two unconditional strategists meet ($1/2\,v - \ m$), so the conditional strategy is the ESS. Note that 'owner wins' is a 'commonsense solution' but the 'paradoxical solution' of intruder wins may also be an ESS. A well known example of the common sense solution is the speckled wood butterfly (Davies 1978). An example of the paradoxical solution is the Mexican social spider (Burgess 1976; Dawkins and Krebs 1978).

The use of the conditional strategy effectively avoids the war of attrition. However, should mistakes occur in the assessment of condition and both contestants assess themselves as being 'owners' then a war of attrition will result (Davies 1978). The assessment required is simple and mistakes are therefore presumably rare.

Ownership influences RHP

In this and the following two sections we discuss situations in which ownership influences some feature that is relevant to the expected pay off. It is common, particularly in fights for access to mates, that the two contestants assume different methods of fighting. These methods are imposed by the positions in which the animals find themselves at the start of the contest. For example, in those species in which males exhibit mate guarding by holding the females it may be physically more difficult to prize the holder from the female than it is to retain the hold on the female. This appears to be the case in dung flies (Sigurjonsdottir and Parker 1981), newts (Verrell 1986) and *Gammarus* (Elwood *et al.* 1987; Ward 1983). Similar arguments apply to scorpion fly contests in which the holder appears to have a positional advantage in holding onto prey items (Thornhill 1984) and to fights over burrows in which the animal within the burrow has an advantage, e.g. in stomatopods (Dingle 1983).

Owners value resources more highly than do intruders

There may be cases in which owners value the resource more highly than do the intruders because the former have already made some prior investment that would have to be made by the intruder if it won the contest. Several examples are given by Parker (1974) and here we will consider just one. In the case of dung flies there is a post-copulatory mate-guarding phase. Should an intruder manage to take over early in this phase he may be able to fertilize 80% of the eggs (Parker 1970). Copulation may thus be regarded as an investment phase and a holder has already made this investment. Should he be ousted the new male has to copulate with the female and guard her against takeovers. Should the holder retain the female he only has to guard her. Thus the pay off for the holder retaining the female is the value of fertilization less the cost of guarding, whereas the pay off for an intruder is the value of fertilization less both the cost of guarding and the cost of copulation. Thus, in a war of attrition, the holder may be able to allocate more time to defence than the intruder can to attack. In the case of dung flies, however, it is difficult to disentangle asymmetry in resource value from positional advantage.

Ownership provides greater information concerning resource value

The fact that an individual has been in possession of a resource may enable more accurate assessment of the resource value. This might occur in the case of ownership and occupation of a burrow, which cannot be entered and assessed by the intruder, as in iguanas (Rand and Rand 1976). Other examples occur in assessment of the quality of sites for webs in the spider *Agelenopsis aperta* (Riechert 1984) and in assessment of the reproductive value of female spiders (Austad 1983). Austad (1983) has calculated changes in the value, to males, of female bowl and doily spiders at different times during the mating sequence. These changes can only be assessed by the holder and not by the intruder. Thus the holder may adjust its fight time to match the assessed value of the resource whereas the intruder will only be able to allocate a fight time appropriate to the average resource value.

For the examples in these last two sections the consequence of the owner:intruder asymmetry is to influence the actual (Owners value resources more highly than do intruders, p. 139) or perceived (this section) resource value in a different manner for the two individuals. Thus the effect is directly comparable to those cases of resource value asymmetry arising without ownership asymmetry and discussed above (Resource value asymmetries, p. 134). In these cases, then, the resource value is the important feature, not ownership *per se*.

CASES OF TWO OR MORE ASYMMETRIES

Examples given above show that almost any asymmetry, be it pay-off relevant or pay-off irrelevant, may be used to settle a dispute. Problems arise in understanding real contests because more than one asymmetry frequently exists. Sometimes two asymmetries, each having a similar effect, may occur. For example, when the owner of a resource also happens to be larger, clearly we can predict a win for the larger owner. But what if the owner is the smaller of the two, then which asymmetry would be used? This problem has been modelled by Hammerstein (1981) and the solutions are complex. However, a general feature emerges from these analyses, and that is to 'respect ownership status if the difference in fighting ability is below a critical value, and respect fighting ability for larger differences'. The critical value is determined more by the cost of injury than by the value of the resource (Hammerstein 1981). One example of this comes from studies of fights of the spider *Agelenopsis aperta* (Riechert 1978). The majority of these fights are won by the larger animal, but in nine out of ten cases in which the weight difference was less than 10% the owner won. In these ten cases it seems that the size difference was not sufficient to be decisive. In scorpion flies also, both ownership and size influence the probability of victory (Thornhill 1984). In this case, however, ownership seemed to confer an advantage in RHP, due to the owner's grip on the resource, and settlement of the dispute simply follows an assessment of total RHP.

Other cases may occur in which both RHP and the resource value are asymmetrical. Again it is easy to predict that an animal with the higher RHP and that values the resource more highly will gain the resource. But what if the animal with the lower RHP values the resource more highly? These situations have also been modelled (Hammerstein and Parker 1982; Parker and Rubenstein 1981) and again the analyses are mathematically complex but the conclusions relatively simple. Of particular interest to the study of hermit crab fights is the conclusion of Hammerstein and Parker (1982) that when more than one pay-off asymmetry exists each animal should be prepared to accumulate costs in a contest until those costs exceed the benefits of winning. A war of attrition may result, in which case the contest ceases when the animal that is prepared to pay only the lower cost gives in. Alternatively, if both animals have perfect information about the asymmetries then the one prepared to pay the lower cost should terminate the contest immediately and thus avoid the cost of fighting up to the point when it would otherwise capitulate. The course of the contest is thus predicted to be determined largely by the accuracy of the information available to the animals, i.e. their assessment abilities. Because of the complexity of the information required, however, it is unlikely that this may be assessed fully prior to the contest and thus assessment may continue during the course of the contest. Animal x should be the winner if:

$$\frac{V_x}{K_x} > \frac{V_y}{K_y}$$

where V = the value of the resource and K = the cost of the contest. That is, animal x will have more units of expendable fitness to allocate to the fight. It is this asymmetric war of attrition with incomplete information that we explore with reference to hermit crab fights, but first we should attempt to define more fully the costs and benefits involved in these encounters.

BENEFITS OF VICTORY IN SHELL FIGHTS

The disputed resource is the defender's shell, which has a certain value (V_a) to the attacker. The value of gaining this shell is dependent not only on its quality but also on that of the attacker's shell. That is, it is the relative quality of the shells or the potential gain from an exchange that is important to the attacker. The attacker has information concerning the quality of its own shell prior to the contest but must gather information concerning the quality of the defender's shell during the early stages of the interaction (Dowds and Elwood 1983). The information gained, however, is not perfect, as demonstrated by the number of crabs that return to their original shell after evicting the defender and testing its shell (Chapter 7). The suitability of a shell is determined not only by its size and weight, which are assessed during the fight, but also by its internal configuration and volume, factors that can only be assessed directly by the crab entering the shell.

The value of victory to the defender (V_d) is more complex. First, there is the value to itself of its own shell relative to that of the attacker's shell. The defender will have good information about its own shell prior to the fight but is unlikely to be able to evaluate the attacker's shell during the contest (Dowds and Elwood 1983; Elwood and Glass 1981; Neil 1985), as the defender spends the duration of the fight withdrawn into its shell. Thus the defender cannot estimate accurately the value of exchanging shells (but see Hazlett 1987 and later in this Chapter). However, if the defender is in a very poor shell it is more likely to gain from an exchange than if it is in a good shell. A complicating factor is that not all shells are taken permanently by the attacker when defenders are evicted, and thus the cost of eviction must take into account the probability that the shell will be taken by the attacker.

There is also a second cost associated with eviction. The evicted defender is without a shell while the attacker makes the decision as to which shell it will inhabit. During this time the defender will have little defence against predation and this risk must be taken into account. Furthermore, if the defender is disturbed at this time by other crabs it may be forced away from the scene of the fight and thus be unable to enter the shell that is left. Indeed, other crabs may take the shell left by the attacker before the defender can take it (P. Abrams pers. comm.), or the shell may be removed by wave action. Thus there is a risk of being naked for some considerable time and/or being left a very poor shell after a number of crabs have made their choice. Thus the value of victory (V_d) to the defender is the avoidance of a variety of costs associated with eviction, in addition to the avoidance of suffering a decrease in shell quality.

COSTS OF FIGHTING

The costs to the attacker (K_a) in continuing the fight are in terms of time and energy. There may also be an increased risk of predation due both to attracting the attention of predators by the actions of the fight, particularly shell rapping, and because the attacker's attention may be concentrated on the fight and hence its vigilance for predators lowered. There seems to be little chance of being wounded by the defender.

The costs to the defender (K_d) include those discussed for the attacker. There is also a small risk of wounding when the attacker pulls at the defender's appendages, as these may be either pulled off or damaged and later autotomized (pers. obs.). It is not clear whether shell rapping imposes a cost on the defender; it may cause disorientation by stimulating the statocysts. Hazlett (1987) suggests that shell rapping communicates aspects of shell quality to the defender and this is discussed in a later section. Being withdrawn into the shell is likely to reduce the efficiency of respiration and this may become an important consideration in extended contests.

HOW IS THE CONTEST SETTLED?

The evolutionarily stable strategy (ESS) for a contest with pay-off asymmetries is for each animal to be prepared to fight until the costs of the fight exceed the benefits of victory (Parker and Rubenstein 1981; Hammerstein and Parker 1982). The winner is thus the animal that can allocate the greatest fight time without the costs exceeding the benefits (Fig. 8.6). However, if the participants have good information about the costs and benefits to each other then it will be to the advantage of the eventual loser to terminate the fight immediately and avoid the cost of a futile fight (Parker and Rubenstein 1981; Hammerstein and Parker 1982). Thus the contests may be settled without combat. If, however, the information is incomplete or insufficient to determine accurately which animal will persist the longer then the contest will be settled by combat. Hammerstein and Parker (1982) provide mathematical formulations for the ESSs of asymmetric contests. They generally assume a linear increase in costs over combat time, but conclude that the ESSs may not be markedly different if there is a non-linear increase in costs. This is important as it is doubtful that costs increase linearly in real animal

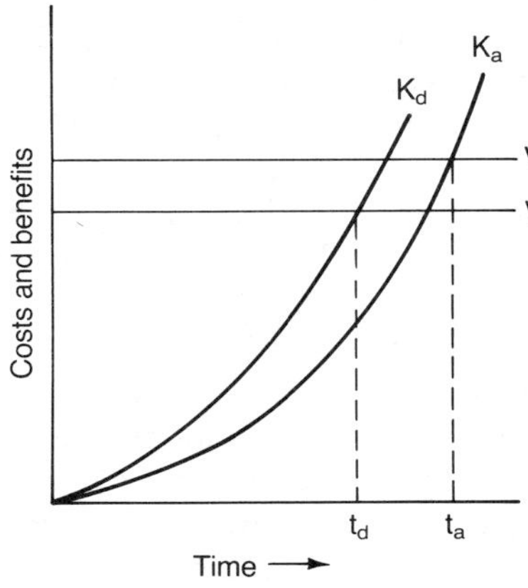

Figure 8.6 Hypothetical comparison in the costs to the attacker (K_a) and defender (K_d) and the benefits of victory to the attacker (V_a) and defender (V_d) show that the attacker may fight for time t_a and the defender for time t_d. The defender will give up at t_d.

contests and we have good cause for doubting that this is the case in hermit crab contests (Elwood and Neil 1986b; Fig. 8.6).

The attacker generally does not escalate the fight immediately so we predict a low cost initially with a much higher cost incurred by both animals when shell rapping commences. If oxygen depletion for the defender is a problem it is likely that the associated costs will increase more rapidly over time and general fatigue for both animals may increase in a similar fashion. Thus we predict that the rate at which costs accumulate will increase during the course of the contest for both animals. At any point in time, however, the costs incurred are likely to be different for the two contestants.

It is predicted that if $V_a/K_a > V_d/K_d$ then the attacker should win, and if the converse is true then the defender should win (Fig. 8.6). With incomplete information the contest should be settled by a war of attrition, in which case the ESS predicts a negatively exponential distribution of fight durations and the distribution of the durations of hermit crab fights (Fig. 8.2) approximates this predicted distribution. This supports the hypothesis that these animals engage in a continuous war of attrition with incomplete information. This model, however, also predicts that the distribution of fight durations should be similar even when there is a 'role asymmetry'. The term 'role asymmetry' is not used in the manner of Hammerstein and Parker (1982) in which the animals identify which has the greater and lesser available fight duration, but rather in the simple sense that there is an attacker and a defender as in fights between dung flies (Sigurjonsdottir and Parker 1981), spiders (Riechert 1978) and scorpion flies (Thornhill 1984). When we compare the distributions of the durations of fights determined by the defenders giving up (i.e. evictions), with those determined by the attackers giving up (i.e. incomplete fights), we see that they are markedly different (Fig. 8.7), and only the latter approximate to a negative exponential distribution. Thus the ESS model of the continuous war of attrition as put forward by Hammerstein and Parker (1982) is not supported by these data.

A closer examination of the distributions of fight durations (Fig. 8.7) allows certain inferences to be made concerning the manner in which these contests are settled. A large number of incomplete fights terminate in the first 60 seconds (73% of incomplete or 42% of all fights), whereas very few of the fights that result in eviction terminate within this period (5% of evictions or 2% of all fights) and these distributions are significantly different ($\chi^2 = 170.9$, df $= 1$, $P < 0.001$). The reason for this appears to be the lack of information that the attacker initially has concerning the value of the defender's shell, i.e. the gain (V_a) that might be made from an exchange. During the early stages of the fight the attacker runs its chelipeds over the external surface and within the aperture of the defender's shell (Dowds and Elwood 1983). Often the defender's shell will be perceived as offering no gain, in which case the attacker will stop, and this results in a large number of short contests which end with the attacker giving up.

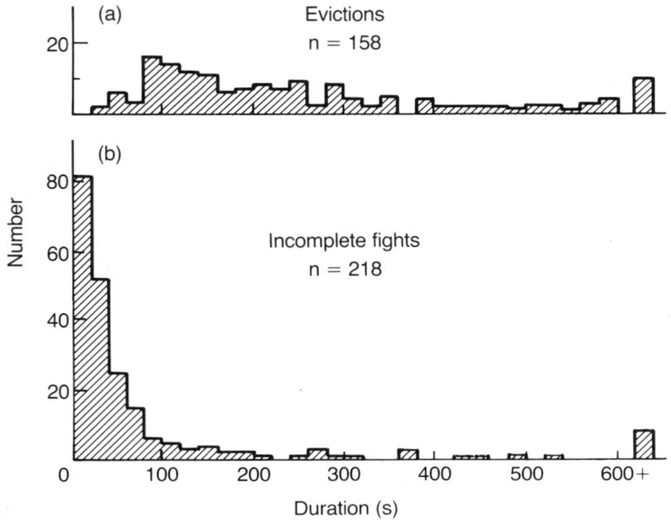

Figure 8.7 Distributions for fights resulting in (a) eviction of the defender and (b) the attacker giving up. Only the latter conforms to a negative exponential. (From Elwood and Neil 1986b.)

DECISION RULES FOR CONTINUING THE FIGHT

What is the defender's best option? Here we assume that in many cases it is not possible for the 'defender' to run away with the resource and thus escape from the contest. In other types of contests with perfect information, costs may be minimized sometimes by playing 0, that is, immediately abandoning the resource. This is not the case in hermit crab contests because abandoning the resource imposes a cost to the 'defender'. Because there is a high probability of the attacker ceasing early, before the costs of defence start to increase at a high rate, the defender's best option is to wait for a short period before leaving the shell. In this manner the defender may avoid the costs of leaving its shell and thus minimize the overall costs of the contest.

The cost of playing time *t* for a defender has two components. First, there is the cost of the fight in time, energy and risk of wounding up to that point. Second, there is the cost associated with eviction multiplied by the probability of being evicted at time *t*, i.e. the probability that the attacker continues the fight at time *t*. If the defender is in an excellent shell the expected cost of eviction is higher than if it is in a poor-quality shell. These costs are represented in Fig. 8.8a which shows a hypothetical curve indicating the increasing cost of the fight. Also shown are three curves representing different costs of being evicted (a = 10, b = 5 and c = 2.5) in arbitrary units of fitness, each multiplied by the probability (derived from the data presented in Fig. 8.7) of the defender being evicted if it plays *t*. These curves are summed and shown in Fig. 8.8b which represent the 'cost curves' for the three eviction costs (three qualities of shell). This representation shows that the minimum cost is achieved by resisting for a certain period and that this period increases as eviction costs increase (i.e. from c to a).

It is also predicted that costs of defence will vary among fights as the relative RHP of the participants varies. This is illustrated in Fig. 8.9a which shows three hypothetical curves representing different costs of defence, with higher costs the smaller the defender is relative to the attacker. We have no measurement of the actual costs to defenders but we have observed injuries to defenders in terms of appendages pulled off and we assume that shell rapping imposes costs to the defender. Thus the assumption that costs to the defender increase as the size difference increases is reasonable. Also shown is a curve representing the probable cost of eviction, i.e. the cost of eviction multiplied by the probability that the attacker will still be in the contest (derived from Fig. 8.7). These are summed in Fig. 8.9b to give three cost curves. These indicate that the contest duration that minimizes cost increases as the cost of defence decreases (i.e., from a to c). Clearly it does not pay the defender to play 0.

What then is the best option for the attacker? We have seen that if the attacker obtains information that no gain may be achieved by taking the defender's shell it effectively plays 0, or as near to 0 as it can achieve given

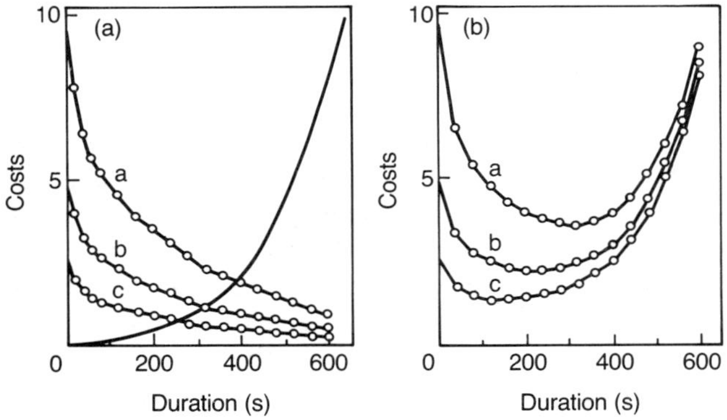

Figure 8.8. (a) Costs for the defender are visualized for three situations for various fight durations. Curves a, b and c represent the cost of being evicted from a good, medium and poor shell respectively, multiplied by the probability of being evicted for a particular defence duration (calculated from Fig. 8.7). The single increasing curve represents the cost of continuing a fight at any one time. (b) These costs are summed to produce 'cost curves' which indicate the fight duration that minimizes costs. (From Elwood and Neil 1986b.)

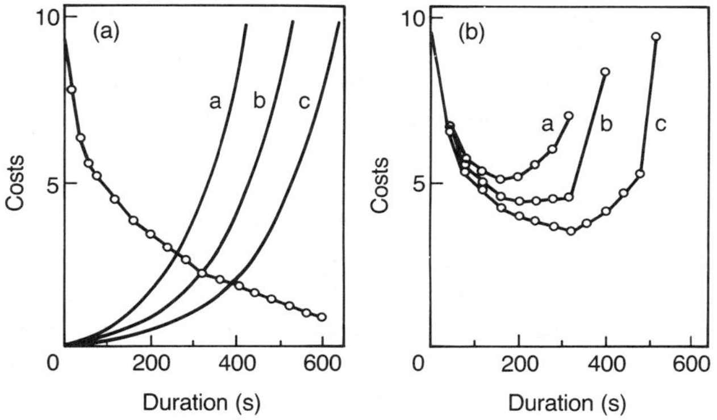

Figure 8.9 (a) Costs for the defender are visualized for various fight durations. The single declining curve is the cost of being evicted multiplied by the probability of eviction at that fight duration (from Fig. 8.7). The three increasing curves a, b and c represent different costs incurred during the fight, i.e. fights with a large, medium and small crab respectively. (b) These costs are summed to produce 'cost curves' which indicate the fight duration that minimizes costs. (From Elwood and Neil 1986b.)

the constraint of having to gather the relevant information about the defender's shell. The question is, for how long should the attacker be prepared to fight if a potential gain is perceived?

We assume that the ESS calculated by Hammerstein and Parker (1982) holds because the attacker cannot estimate for how long the defender will fight, i.e. there is incomplete information. Thus the attacker is prepared to fight until $K_a = V_a$. That is, the attacker will fight for a predetermined period set after gathering information about the resource but prior to giving up, leaving the defender in the defender's shell. Hammerstein and Parker (1982) predict that most contests will be short in this situation and that the distribution of fight durations will follow a negative exponential. This is found in our data for fight durations determined by the attacker giving up (Fig. 8.7b).

TESTING THE THEORY

The theory on decision rules results in four clear predictions, each of which may be tested by experiment.

Prediction 1. As the cost of eviction increases the defender should show increasing resistance to eviction (Fig. 8.8b). The fights that end in evictions are those in which the durations are determined by the defender; the defender decides when to leave the shell and hence when the fight ends. It is predicted that when the cost of eviction is high these fights should be longer and, as a consequence, that fewer fights overall should end in eviction because the attacker is more likely to have given up.

This prediction may be tested in a number of ways. First, we may examine the effect of the quality of the shell that the defender possesses as this has a marked effect on the cost of eviction. To isolate this as a meaningful variable we must be reasonably certain that variation in the quality of the defender's shell is either not detected or is irrelevant to the attacker. This may be achieved by comparing the behaviour of defenders in normal shells with those in shells that have had their apertures partially blocked with resin (Neil 1985). As defenders in the shells with partially blocked apertures are unable to withdraw completely it is difficult for the attacker to detect the blockage until after the defender has been evicted. The cost to the defender in leaving these shells will be less than that of leaving normal shells, since most alternative shells would offer an improvement. That this is the case is supported by the greater probability of eviction of defenders in the modified shells. The fight durations, however, did not differ significantly (Neil 1985).

In fights between naked attackers and housed defenders the cost of being evicted is very high. Although previous studies (Dowds and Elwood 1983) have suggested that defenders could not assess the quality of the attacker's shell it seems reasonable to assume that they can assess whether or not the attacker has a shell. When such fights are examined it is apparent that defenders resist for considerably longer when the attackers are naked than when the attackers are housed (Dowds and Elwood 1985; Elwood and Glass 1981). However, although these findings are congruent with the above prediction, it should be noted that naked crabs may be less efficient at fighting, that is, their RHP may be reduced. This is especially true for shell rapping, a major contest activity. If this is the case it would also be likely to increase defence time (Fig. 8.9).

Another example of higher than usual costs of being evicted occurs with egg-carrying (berried) females. Eggs are carried on the pleopods (abdominal appendages) and eggs may become dislodged when a crab is evicted from the shell (Neil and Elwood 1985). Thus berried females face higher eviction costs than do non-berried females or males. It is interesting to note that the duration of fights leading to the eviction of berried females are longer than for males or non-berried females (Dowds 1983). Furthermore, berried females defend more vigorously than do males against naked male attackers (Neil and Elwood 1985).

These findings provide support for the first prediction. However, in a study in which crabs in a preferred species of shell (*Littorina obtusata*) competed with crabs in unpreferred species (*Gibbula* spp.) defenders in the former did not defend for longer than did those in the latter. Those in the preferred species, however, were more likely to engage in cheliped flicking than were those in the unpreferred species which suggests that to examine only durations of fights may be an oversimplification of the analysis of cost (Dowds and Elwood 1983). In general, however, we may conclude that there is reasonable agreement with the prediction that the higher the cost of eviction the longer the defenders will resist and the fewer will be evicted.

Prediction 2. As the cost per unit of time of fighting increases for the defender the duration of defence should decrease and more evictions should result (Fig. 8.9b).

This prediction may be tested by comparing fights that vary with regard to the relative RHP of the participants. Size is frequently used to assess RHP and therefore we may examine fights in which individuals differ in size. Such data are available for fights involving naked attackers (Dowds and Elwood 1985; Elwood and Glass 1981) and for housed attackers (Dowds and Elwood 1985; Neil 1985), and in each study an increase in the size difference resulted in an increase in the number of evictions (Chapter 7). Furthermore, if the defender is lacking a major cheliped, and thus presumably having a reduced RHP, it is more likely to be evicted (Neil 1985). However, although the prediction concerning the probability of evictions is upheld none of the differences in fight durations were significant. Experiments on *P. gaminus* involving some attackers having their chelipeds covered by a plastic tube, and hence preventing their use in fighting, has demonstrated that defenders persist longer when the attacker has no ability to grasp the defender's appendages (Imafuku 1989).

A study in which the size difference of the crabs was a variable, as was the cost to the defender of eviction (each crab was in the other's preferred shell) found significant negative correlations between duration of the fight and the size difference (Dowds 1983). Furthermore, significant negative correlations were noted for cheliped flicking, with defenders much smaller than attackers less likely to perform this activity than were defenders closer to the size of their opponents. Thus, although two variables are confounded (RHP and costs of eviction) the findings are entirely consistent with the first and second predictions.

Prediction 3. The greater the potential gain to the attacker the longer it should be prepared to fight (Fig. 8.6). This should result in a greater proportion of fights won by the attacker and longer incomplete fights (in which we can assess the time allocated by the attacker) for attackers with much to gain from an exchange.

Attackers housed in unpreferred shell species win more fights compared to attackers in preferred shell species and they win an even higher proportion of fights if the opponent is in a good-quality shell (Dowds and Elwood 1983). Similar findings have been noted for attackers in shells with the apertures partially blocked (Neil 1985). In both cases the attacker is more likely to win if the potential gain (improved shell quality) is high (Chapter 7). However, differences in the time allocated to fights by attackers as a function of their potential gain can not be examined because very few fights end without victory to the attacker (i.e. are incomplete) if it can make a substantial gain. In order to analyse fight duration we can examine the data for naked attackers, which clearly have a greater potential gain. As predicted, they engage in longer fights prior to giving up than do housed attackers (Elwood and Glass 1981; Dowds and Elwood 1985). Furthermore, berried female attackers, housed in shells that are too small and hence likely to damage their eggs, are more likely to evict a defender (Dowds 1983). However, once again there were insufficient numbers of incomplete fights to test whether or not the berried female attackers were fighting for longer periods. If we examine complete fights, i.e. those determined by the defender giving up, then these are longer when the attacker is berried than unberried. Thus although these data do not strictly tell us about the fight durations set by the attacker they do suggest that the greater success of berried attackers is due to their greater persistence in fights. There is thus evidence to support both parts of the prediction.

Prediction 4. The greater the cost of fighting to the attacker the shorter will be the incomplete fights, and the greater will be the proportion that end without eviction (Fig. 8.6).

This may be tested by comparing fights that differ with respect to the relative RHP of the participants. The data from such comparisons clearly support the prediction of more wins by the attackers when they are relatively larger than the defenders (Dowds and Elwood 1985; Elwood and Glass 1981; Neil 1985). Furthermore, the data for incomplete fights show that they are considerably longer when the attacker is relatively large (Dowds and Elwood 1985). Thus both parts of the prediction are clearly upheld.

CONCLUSION ON DECISION RULES

Each of the four predictions had two parts. One concerned the probability of the attacker being victorious and for all four predictions this was clearly upheld. However, the main point of the analysis was to make predictions concerning the duration of the fights and for this the data were less clear. Data that support predictions that defenders increase their fight time according to the potential cost of eviction (prediction 1) or cost of continuing the fight (prediction 2) may have alternative explanations. The data that support predictions that attackers should persist for a longer time when potential benefits are great (prediction 3) and for a shorter time if costs are great (prediction 4) are unequivocal.

Thus, overall, six of the eight components of the predictions are clearly upheld and for the other two there are congruent data which may have alternative explanations.

NEGOTIATION OR AGGRESSION?

Hazlett (1978, 1983, 1989) has proposed an entirely different model to explain these interactions. He suggests that they are not of an aggressive nature but rather are a process of 'negotiation'. Both crabs are viewed as exchanging information about the qualities of their respective shells and an exchange is determined primarily by the 'defender'. It is proposed that if a defender may gain, an exchange is more likely to occur (Hazlett 1983). The model is explicit in its predictions but does not consider the potential costs to the defender when naked or, alternatively, the costs of continuing the interaction. Hazlett (1983) makes two main predictions of the model:

Prediction 1. As the shell quality of the defender declines for the defender the more likely it becomes that it will evacuate the shell.

This is similar in many respects (but for different reasons) to prediction 1 of the aggression model (above) and, in general, the prediction is upheld. It is clear that defenders have good information concerning the quality of the shell in their possession and that this information influences their behaviour.

Prediction 2. As the quality of the attacker's shell increases in terms of suitability for the defender an exchange becomes more likely.

It has been shown that attackers in poor-quality shells are more likely to effect an exchange than if they are in good-quality shells (Dowds and Elwood 1983; Neil 1985). This occurs when the shells are of an unsuitable species (Dowds and Elwood 1983) or if the apertures have been partially blocked with resin (Neil 1985) and thus they are unsuitable also for the defender. Furthermore, in a specific test of the negotiation model two groups of crabs were tested. In one group the attacker would have the greater gain from an exchange and in the other the defender would have the greater gain (Elwood and Glass 1981). In the former group the larger crab was given a shell 10% lighter, and hence smaller, than that normally preferred by the smaller crab. The smaller crab was given a shell of the preferred weight of the larger crab. In the latter group the larger crab was given the preferred shell of the smaller crab and the smaller crab was given a shell that was 10% too heavy for the larger crab. Thus, although both crabs in both groups could gain from an exchange, only in the first group would the attacker obtain its preferred shell, whereas only in the second group would the defender obtain its preferred shell. Thus the negotiation model would predict more changes in the second group. The results were the opposite; that is, there were more changes when there was more gain to the attacker.

Finally, we have seen above that, although naked attackers take longer to evict defenders they do eventually obtain the shell. It is difficult to see how the negotiation model can explain this as the defenders suffer a considerable cost (Elwood and Glass 1981). It is clear that data collected in our laboratory contradict those of Hazlett (1978, 1983) concerning the negotiation hypothesis.

One of the key differences between the aggression and negotiation hypotheses concerns the assessment abilities of the defender. The negotiation model suggests that the defender can assess the attacker's shell. Our data indicate that it cannot. More recently Hazlett (1987) has suggested that *Clibanarius antillensis* can gather this information by means of determining the fundamental frequency of the tapping sound made during shell rapping. It is suggested that this frequency is related to shell volume and that the crab is able to discriminate the frequency from its own shell from that of its opponent's shell. Shell rapping is thus viewed as a mechanism by which the attacker provides information for the defender. Having gained this information the defender is able to make decisions based on the actual gain (or cost) of an exchange rather than the average expected gain or cost as outlined above (Fig. 8.8). How this altruistic behaviour could have evolved is not clear, as the evolution of signals for the benefit of non-relatives seems to be unlikely (Dawkins and Krebs 1978). Of course, shell rapping might provide additional information for the attacker about the defender's shell, but this would be redundant as the attacker assesses the defender's shell with its chelipeds. Shell rapping thus seems unlikely to have evolved for that purpose. It seems more likely that it has an aggressive function, either communicating aspects of the attacker's RHP, inflicting some damage to the defender or threatening to do so, but should it contain information about shell quality, it would certainly pay a defender to assess that information. This assessment may require learning from previous encounters and this could explain why older crabs appear to fit the negotiation model of Hazlett more convincingly than do young crabs (Elwood and Glass 1981; Hazlett 1983, 1987). If it were disadvantageous to the attacker to allow the defender to get information from shell rapping (if there is any to be had), as it appears to be, then selection should disfavour this activity, unless the advantages outweigh the disadvantages.

In order to test the negotiation model Hazlett (1987) observed interactions between *Clibanarius antillensis* individuals and noted their outcome. He then removed the crabs from their shells and modified the attackers' original shells to either (i) suit the defenders more if a shell exchange had not occurred, or (ii) suit the defenders less if a shell exchange had occurred. The crabs were then placed in their original shell and allowed to interact again. Altering the weights of the attacker's shells had little effect, but changing the internal volumes yielded a significant tendency to alter the outcome between the two encounters. That is, those that had not changed before now were more likely to do so and those that had changed before now were less likely to change. Hazlett suggested this to be due entirely to motivational change on the part of the defenders, i.e. those whose shells were not altered. However, grinding down or adding to the interior of a shell is bound to alter the attacker's perception of its own shell. Indeed, about 15% of the original attackers did not attack in the second encounter; clearly their motivation had changed. Thus it is probable that at least some of the altered outcomes of encounters were due to changes in the motivational state of the attackers, although this cannot account for all the altered outcomes. These data thus support the negotiation model as they suggest that, at least in some cases, the defender can obtain information about the attacker's shell.

The question still remains whether negotiation is a better concept than aggression for describing hermit crab shell fights. Hazlett (1987) contrasts the two models but his view of the aggression model is simplistic in that he assumes that a larger crab should always win. It is not surprising that relatively few fights fit this model as it does not take into account either the degree of size difference or the amount of gain an attacker may make in an exchange. Hazlett (1987) assumes that all attackers that commence shell rapping have assessed a potential gain, but our data show that assessment continues throughout a fight and that mistakes occur, with the attacker returning to its original shell even after moving into that of the defender. Furthermore, in comparing the two models all those fights in which the attacker assesses a potential loss and therefore does not escalate (shell rap) are omitted in the studies of Hazlett (1983, 1987), thus further biasing the data set towards the negotiation model, in which the defender plays a more important decision making role.

These are serious criticisms of the methods used in those experiments that suggest negotiation, but they do not by themselves discount the possibility that defenders can learn to obtain information about attackers' shells. However, do we need an entirely new model of negotiation to explain such a phenomenon? Clearly the additional information will allow a defender to alter its tactics and this would result in more exchanges when the cost to the defender is low and fewer exchanges when the cost is high. This, however, does not alter the predictions of the war of attrition outlined above – indeed, it is entirely congruent with that model; that is, if $V_a/K_a < V_d/K_d$ then the defender should win. All the negotiation model implies is that the estimate of V_d is more accurate.

Hazlett does not suggest that all hermit crab shell fights are a process of negotiation. His studies on *Calcinus tibicen* failed to support the negotiation hypothesis (Hazlett 1984) and further studies suggested that interactions between these crabs might involve aspects of aggression as well as negotiation (Hazlett 1989). It was noted, however, that the number of shell raps executed by the initiating crab (attacker) was higher in cases that resulted in an exchange than for non-exchanges (mean of 57 versus 24; $t = 3.45$; $P < 0.01$). Clearly, this suggests that the motivation of the attacker is an important determinant of the outcome of these interactions. How the motivational state of the attacker and defender influence the duration and outcome of fights in *P. bernhardus* is considered in the final chapter.

To conclude on the negotiation model we may again consider the two main predictions. One, similar to a prediction of the aggression model, is upheld. The other, dissimilar from the aggression model, has been tested in several experiments in our laboratory and the data have been unequivocally against the negotiation model.

SUMMARY

Theoretical and experimental studies concerning durations of fights are reviewed. In particular, the role of asymmetries is considered. Predictions from the asymmetric war of attrition model (Parker and Rubenstein 1981; Hammerstein and Parker 1982) are compared with observations on hermit crab fights but the data are not fully compatible with that model. The discrepancy appears to be due to the costs to the defender of abandoning the shell early in a fight; defenders may minimize costs by fighting for a time greater than zero. A model is developed to express the costs and benefits of attackers and defenders and predictions of this model, concerning the effects of resource value and the relative RHP on fight durations, are congruent with data. The data do not support the suggestion that hermit crabs negotiate over shells instead of competing for them by aggression (Hazlett 1978).

CHAPTER 9

Motivational change during aggressive encounters

It is the purpose of this chapter to examine motivational processes that control aggressive encounters of hermit crabs and other species, particularly those in which fight duration decides the winner. This type of encounter has been considered extensively from a functional point of view in the previous chapter. Various evolutionarily stable strategies have been calculated (Parker 1974; Parker and Rubenstein 1981; Hammerstein and Parker 1982) that vary with the assessment abilities of the animals. Without any assessment the animals are predicted to fight with animal A winning if $V_a/K_a > V_b/K_b$ (where V = value of resource and K = costs of fighting) and the fight ending when B gives up at the point when its accrued costs equal the expected value of the resource (Parker 1984). If perfect assessment of opponents and resources is possible, however, fight times will be zero because the contestant with the lower V/K will avoid incurring unnecessary costs. It is likely, however, that few fights conform to these simple models. It is more likely that information is gathered during encounters that enables decisions to be made which determine the winner (Enquist and Leimar 1983, 1987; Parker 1984). Thus an animal will have to enter the contest in order to obtain information but that information will enable the animal to cease fighting when it estimates that the costs are likely to exceed the probable benefits. The mathematics utilized in calculating the ESSs of such encounters is complex (e.g. Hammerstein and Parker 1982; Enquist and Leimar 1983) due to variation in the mode of acquisition of information, and in the way in which the information is used.

Although calculating the ESS for these encounters is complex this need not mean that the motivational processes that control the behaviour are complex. In this chapter a relatively simple two-dimensional state space model (Chapter 1) is presented that generates fight durations similar to those observed in real fights. This model is then compared to other recent models of motivational change during aggressive encounters. The key feature of the present model is that information concerning the value of the resource (an estimate of V) and of the likely cost of the fight (an estimate of K) are gathered during the contest and cause a change in the animal's motivational state. We attempt to model these changes in the estimate of V and K during the contest. Decisions made during the fight depend upon the motivational state, which is examined in a way that attempts to integrate both causal and functional models.

RESOURCE ASSESSMENT AND MOTIVATIONAL CHANGE

The way in which asymmetries in resource value influence the ESS has been explored theoretically (Bishop *et al.* 1978; Enquist and Leimar 1987) but there are few empirical studies (e.g. Rand and Rand 1976; Austad 1983) and these usually concern differences in information available to the contestants rather than differing resource value *per se*. For example, a spider that has transferred sperm for x minutes should be more certain of the value (in terms of unfertilized eggs) of retaining the female than is a newcomer, which should assess the female as of 'average value' (Austad 1983).

In the case of hermit crab fights there may be real differences in the value of particular shells to each of the participants (Hazlett 1978; Elwood and Glass 1981). Studies of *Pagurus bernhardus* have shown that when an attacking crab initiates a contest it has little or no prior information as to the quality of the opponent's shell and that this information is gathered during the contest (Dowds and Elwood 1983). The attacker feels the exterior of the shell and then turns the shell, still with the opponent withdrawn into it, and feels the aperture (Dowds and Elwood 1983) and it is these acts that appear to be used in information gathering (Chapter 7).

There is a particular advantage of using hermit crabs in the study of resource assessment during aggressive encounters. They perform the same acts and gather the same information about shell quality in non-aggressive situations (Elwood and Stewart 1985; Neil and Elwood 1986). Studies on hermit crabs have been conducted specifically to examine motivational change during resource assessment without the confounding effects of an opponent (Elwood and Neil 1986a; Jackson and Elwood 1989b, 1990).

After assessing an empty shell a hermit crab may enter it or reject it (Chapter 5). Shells that offer an increase in shell quality are more likely to be entered and are entered after a shorter investigation period (Elwood and Neil 1986a; Jackson and Elwood 1989b). A short investigation might occur because (i) the shell that the crab is currently occupying is of low value, or (ii) the new shell may be of a more preferred size or species (Jackson and Elwood 1989b). It has been demonstrated that crabs gather information concerning shell quality from assessment of the exterior as well as the aperture (Jackson and Elwood 1989b).

The differences in the timing of decisions concerning unoccupied shells have been explained as being mediated by changing levels of causal factors for shell acquisition (Chapter 5). In state space terminology (McFarland and Houston 1981) the assessment may be viewed as changing the position of a point in motivational space along a single vector, i.e. within a single dimension. The level of causal factors for shell acquisition will vary over time as the crab gathers information concerning the shell being investigated (Elwood and Neil 1986a; Jackson and Elwood 1989b).

If a crab encounters a shell that has the aperture blocked, that shell may be investigated and then rejected. A model presented by Elwood and Neil (1986a) and Jackson and Elwood (1989b) suggests that the level of causal factors will drop during the investigation to a threshold or switching line for rejection and the crab will cease investigation at that time (Chapter 5). The investigation of a blocked shell is analogous to the situation of an attacker examining a shell occupied by a persistent defender. In both cases the shell cannot be entered by the investigating crab. Thus studies of motivational changes during investigation of artificially blocked shells may offer insights into motivational changes during shell fights.

Neil and Elwood (1986) induced crabs to occupy shells of 25%, 50%, 75% or 100% of their preferred shell weight and then offered these crabs either a very poor shell (25% preferred weight) or a very good shell (100% preferred weight) with their apertures blocked with resin. All these shells were eventually rejected but there were great differences in the times for which crabs persisted in investigation (Fig. 5.9). Those crabs housed in the poorest shells showed the longest investigation times and those offered preferred shells also persisted for a long time (Fig. 5.9). It has been demonstrated by the use of interruption techniques that crabs in poor shells have a higher level of causal factors for shell acquisition at the start of shell investigation than do those in good shells (Jackson and Elwood 1990). Thus the four qualities of shells (25%, 50%, 75% or 100% of preferred weight) occupied by crabs in the experiment of Neil and Elwood (1986) (Fig. 5.9) will result in different levels of causal factors at the start of investigation. The poorer the occupied shell, the higher will be the initial level of causal factors (Fig. 9.1). It has been demonstrated that during external investigation information is gained that alters the causal factors for shell acquisition (Jackson and Elwood 1989b). Thus crabs that examine the exteriors of preferred shells will perceive that they are of preferred size and show an increase in causal factors, whereas those that examine shells that are too small will show an immediate decrease. When a preferred shell is turned, however, and the blocked aperture is detected the causal factors will decline. For the sake of simplicity we assume that at this stage the rate of decline is the same irrespective of the starting level of causal factors. This assumption allows the plotting of trajectories that cross the switching line for rejection (Fig. 9.1) at the times determined by experimentation (Fig. 5.9). We have also made the simplifying

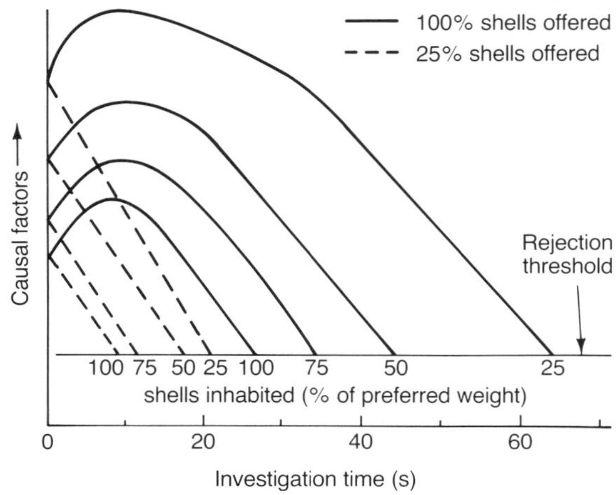

Figure 9.1 Trajectories of levels of causal factors for crabs housed in four qualities of shell investigating shells either of the preferred size or too small, which are all blocked.

assumption that the level of causal factors at the start of investigation is set entirely by the occupied shell. Previous experiments (Chapter 5), however, have suggested that some knowledge of the shell to be investigated is gained by visual inspection during approach. Thus crabs approaching very small shells (25% of the preferred weight) are likely to have a lower level of causal factors for shell acquisition than are those approaching preferred shells. These trajectories represent changes in motivational state when investigating blocked shells and, as stated above, this situation is analogous to that of an attacking crab in a shell fight. In other words, similar trajectories may represent changes in the attacker's estimate of the value of the defender's shell, i.e. V.

Let us discuss the two extreme trajectories in more detail and plot the changes in motivational state at 5-second intervals (Fig. 9.2). Because we are now discussing changes in motivational state during fights we will refer to the vertical axis as 'estimate of resource value (V)' so as to make the causal model congruent with functional models of aggression. An arbitrary scale may also be placed on the vertical axis. Trajectory 'a' represents the changes in the estimate of the value of the opponent's shell when the attacking crab is in a poor shell and the defender is in a good shell. Contrast this with trajectory 'b', which represents the changes in the estimate of the value of the opponent's shell when the attacker is in a good shell and the defender is in a poor shell. Note that in both cases it is assumed that the defender persists in the contest. It is clear that these trajectories differ greatly from case to case and, if the estimate of V varies so widely, so might its influence on fight duration.

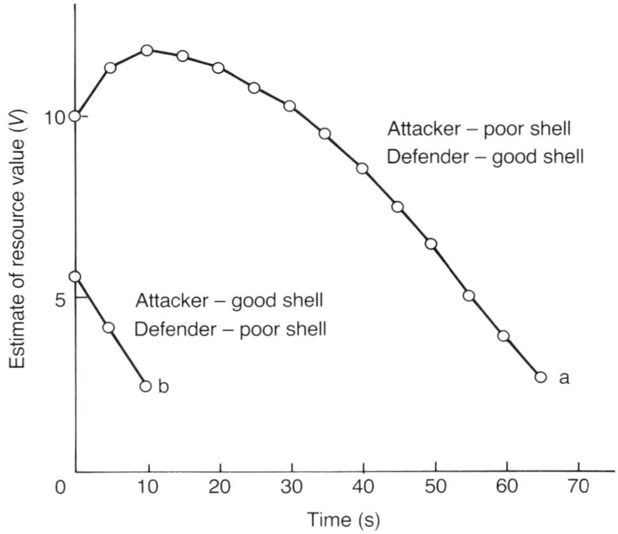

Figure 9.2 The two extreme trajectories from Fig 9.1 are plotted and 5-second intervals are added to these plots. See text for details.

RHP ASSESSMENT, THE COST OF FIGHTING AND MOTIVATIONAL CHANGE

During the initial stages of an aggressive encounter the exchange of cheliped displays enables each crab to assess the relative RHP of the other (Dowds and Elwood 1983, 1985; Neil 1985). The larger crab, or at least the one with the larger chelipeds (Neil 1985), usually initiates the fight and assumes the role of attacker. This initial assessment, however, will change as further information is gained during the fight. For example, when the defender's shell is turned over the defender becomes visible to the attacker; in cases in which small crabs attack larger crabs without an initial display (i.e. attack from behind), fights usually terminate at this stage (Dowds and Elwood 1983). This suggests a reassessment of relative RHP in these cases. It is reasonable to assume that, in the more usual cases of a larger crab attacking after an initial display, assessment will also continue during the fight and perhaps confirm or enhance the attacker's perception of superiority in RHP (Enquist and Leimar 1983). Assessment will continue throughout the fight, not only visually but by tactile means during the grappling sessions which ensue. An assessment strategy implies that the animal attempts to predict the cost of a fight (without actually incurring that cost) (Parker 1984) so that the information gained will alter motivational state.

The expected cost of a fight to the attacker will be determined by a combination of relative RHP and expected fight duration. The estimate of fight duration can be taken from a probability distribution of fight durations (Fig. 8.7). What then is the average fight time that will result in victory to the attacker? This can be calculated as the median of the fight durations that result in eviction (Fig. 8.7a) and equals 204 seconds, i.e. if the attacker persists for this time it will result in victory on 50% of the occasions. The expected fight duration will change as the fight progresses but will always be the median of the fight durations that result in eviction that are greater than the current fight time. It is thus clear that the expected fight duration increases as the fight progresses so the expected costs will increase (Fig. 9.3). Thus after the elapse of 100 seconds fighting the expected fight duration will have increased to 240 seconds (Fig. 9.3). We have suggested (Elwood and Neil 1986b) that actual costs do not rise linearly; rather they start low but increase markedly when the fight is escalated (Chapter 8). For the sake of simplicity, however, let us assume that the cost of a fight rises linearly. Thus the increase in the expected cost of a fight when the actual fight time goes from 0 to 100 seconds (Fig. 9.3) is calculated as 240/204 = 1.18. Hence, the expected costs to an attacker will rise by only 18% after the elapse of 100 seconds in the contest.

The other feature that will markedly affect costs of fighting is the relative RHP of the contestants (Dowds and Elwood 1985; Neil 1985). If the defender is about the same size as the attacker the latter will probably have to pay higher costs than if the defender is considerably smaller. We do not know, however, how much higher those costs will actually be. What we can say is that small crabs are virtually never successful in attacks against larger crabs so presumably the costs of such an encounter far exceed the potential benefits of victory. In the calculations of estimates of the value of the resource for

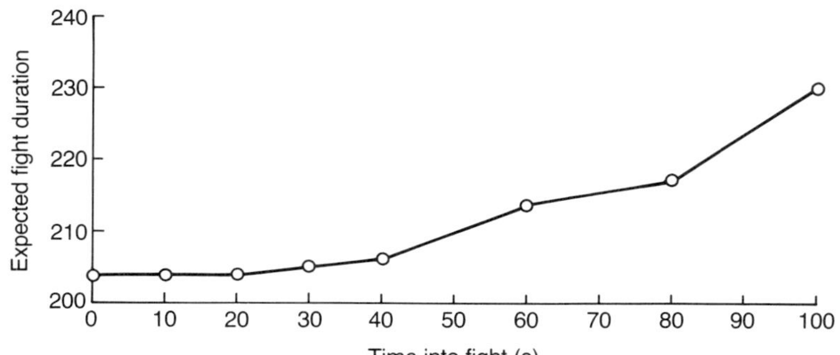

Figure 9.3 Expected fight durations for 50% chance of victory for the attacker at different actual persistence times.

the attacker (Fig. 9.2) the maximum initial value was set at the arbitrary value of 10. Thus the estimated costs to a smaller attacker must be higher than this value. Let us examine two hypothetical cases of the more usual situation of the attacker being the larger crab. In one case the initial expected cost to the attacker is 5 arbitrary units (i.e. the attacker is not much larger than the defender), in the other it is 2.5 arbitrary units (i.e. the attacker is much larger than the defender), but both increase with the passage of time in the fight for the reasons outlined above (Fig. 9.4).

One final point should be considered about the changes in the estimate of costs shown in Fig. 9.4. For each level of initial cost the same expected duration has been calculated. It is expected, however, that relative RHP influences fight duration (Elwood and Neil 1986b). Thus an attacker that is much larger than its opponent would not only face lower costs per unit of time fighting than would one which is not much larger than its opponent, but it could also expect a shorter fight. For simplicity this has been ignored in determining the shape of the lines in Fig. 9.4.

TWO-DIMENSIONAL STATE SPACE

Functional models of fights predict that the important features that determine fight durations are the value of the resource (V) and the likely cost of the fight (K) as assessed by each contestant. The above analysis has indicated how each of these may vary over time for an attacker in a hermit crab fight and, more importantly for the present discussion, how the motivational state changes. Note that for each estimate (Figs 9.2 and 9.4) the time base is indicated at 5-second intervals. What is required is some manner of

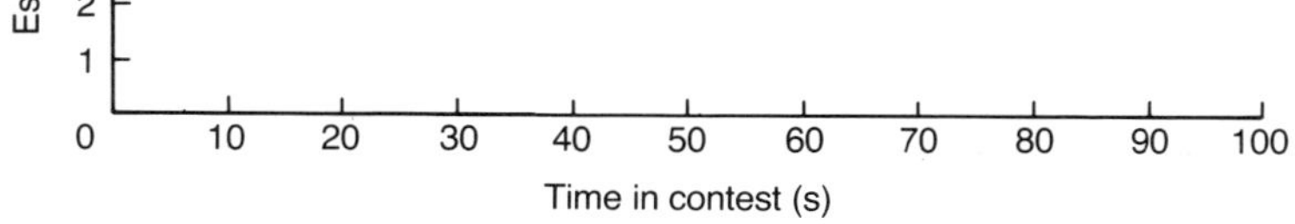

Figure 9.4 The estimate of the cost to the attacker changes during the contest but only by a small amount; (c) is the estimate of an attacker not much larger than the defender whereas (d) is for an attacker much larger than the defender. See text for details.

bringing these two changing estimates into a single motivational state space (McFarland and Houston 1981). To achieve this the estimates may be mapped onto a two-dimensional state space, the axes of which are V and K in arbitrary units (Fig. 9.5). Let us examine the mapping of the estimate of a potential gain in shell quality (Fig. 9.2a) with the estimate of low cost of fight (Fig. 9.4d) to form the trajectory (ad) in Fig. 9.5. At time zero the estimate of V is 10 units, the estimate of K is 2.5 units and these form the first point in trajectory (ad) in Fig. 9.5 (the first point is the filled circle). After 5 seconds the estimate of V has changed to $V = 11.5$ (Fig. 9.2a) but K is still 2.5 (Fig. 9.4d) and these form the second point. By repeatedly mapping at 5-second intervals a trajectory is obtained with a time base.

Theory predicts that an animal should only fight if the estimate of V is greater than the estimate of K. A switching line or motivational dominance boundary may be drawn in the state space, so that for all points above the line $V > K$, whereas for all those below the line $V < K$. The animal should continue the fight for as long as $V > K$, i.e. for as long as the points remain

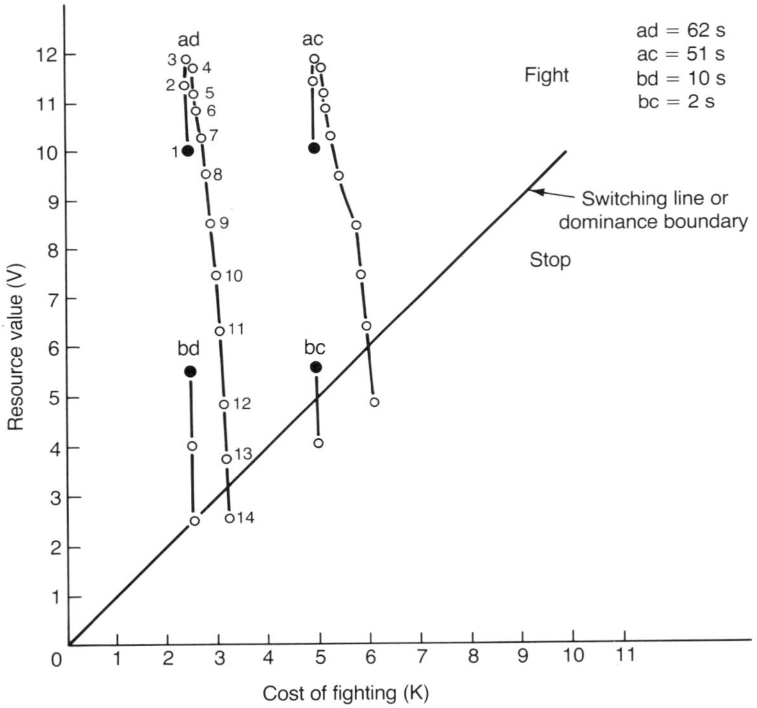

Figure 9.5 Four trajectories of motivational states of attackers are plotted with adjacent points being separated by a 5-second interval. The closed points indicate the point in motivational state at the start of the fight.

above the switching line. As the trajectory crosses this line the crab should cease fighting. By counting the 5-second periods required to reach the switching line we may estimate the duration of the fight. When this is done for the present example of high *V* and low *K* the predicted fight time is 62 seconds.

If *V* is high (Fig. 9.2a) and the cost (*K*) is also high (Fig. 9.4c), we get trajectory ac in Fig. 9.5 which predicts a fight time of 51 seconds. If *V* is low (Fig. 9.2b) and *K* is low (Fig. 9.4d) we get trajectory bd in Fig. 9.5 which predicts a fight duration of 10 seconds. If *V* is low (Fig. 9.2b) and *K* is high (Fig. 9.4c) we get trajectory bc in Fig. 9.5 which predicts a fight of 2 seconds. Note that these are not actual fight durations but the durations for which the attacker should be prepared to fight. In all cases in which the initial estimate of *K* is greater than that of *V* the animal is predicted not to attack; it might, however, be attacked by the other crab. If both crabs estimate $K > V$ then there will be no fight.

MOTIVATIONAL CHANGE IN THE DEFENDER

In real contests the majority of fight durations determined by the attacker, i.e. those contests in which the attacker gave up, are short and the distribution conforms to a negative exponential (Elwood and Neil 1986b, Fig. 8.7b). The distribution of fight durations determined by the defender, i.e. those contests in which the defender gave up, however, is quite different (Fig. 8.7a). There are no very short fights in this group and functionally this has been explained by two factors: (i) the high cost of eviction, and (ii) the high probability of the attacker giving up early in the fight (Elwood and Neil 1986b). Thus, even if the attacker is considerably larger, defenders may minimize costs by fighting for a short duration rather than zero. The question posed here is how may this be accounted for in the state space model?

In hermit crab fights the two contestants assume different roles, and this has a number of consequences of importance in understanding the estimation of *V* and *K*. One major consequence relates to the information that is available to the defender regarding the value of the resource. Clearly the value of retaining one shell depends to a large extent on the quality of the alternative shell. The defender, however, is not in a position to gather information about the attacker's shell because it is withdrawn into its own shell (Dowds and Elwood 1983). Thus the defender will normally only have information about its own shell and the quality of this shell seems to be a major determinant of the defender's persistence. The estimate of *V* is thus likely to remain virtually the same for the defender throughout the fight as no further information is available to alter that estimate. In some species of hermit crab, however, it is possible that a defender may be able to gain further information about the attacker's shell (Hazlett 1987) and in those cases the estimate of *V* will change accordingly. This does not, however, alter the basic motivational model.

The defender will be able to make an assessment of relative RHP and hence estimate *K* from the initial cheliped display, from visual information during the initial stages of the fight, from tactile information while grappling and shell rapping and from the persistence of the attacker. The estimate of *K* will be based on the average fight duration required to retain the shell and this may be derived from the probability distribution in Fig. 8.7b. Note that because most fights end after a short period with the attacker giving up, the initial expected fight duration for the defender is only 28 seconds (i.e. for a cost of 28 seconds the defender can expect to win 50% of fights). The expected fight duration will change as the fight progresses and can be calculated as the median duration of fights resulting in successful defences that are greater than the current fight time. Thus after 20 seconds of the fight has elapsed the expected fight time for the defender will almost double to 55 seconds. If we make the same simplifying assumption as made for the attacker, i.e. that cost increases linearly with time, the increase in the estimate as the fight progresses may be plotted (Fig. 9.6). Two plots are shown in Fig. 9.6, one for a high and one for a low initial estimate of cost; this represents a large size difference and small size difference between the defender and attacker respectively. The key point to note is that at the start of the fight a defender will estimate that a short fight is likely to be successful (28 seconds) whereas an attacker will predict a considerably longer fight if it is to gain the shell (204 seconds). It is thus reasonable to assume that the initial estimate of costs is lower for the defender than for the attacker.

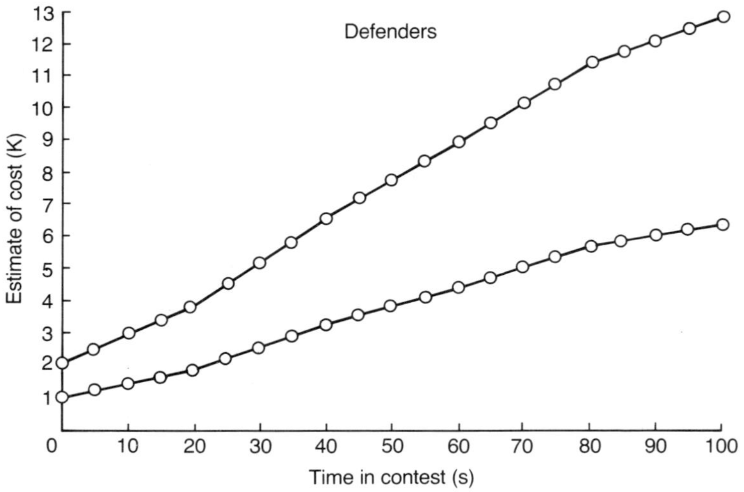

Figure 9.6 Expected costs increase for defenders as the fight progresses. Contrast with the minor increase in expected cost for attackers (Fig 9.4).

Whereas estimated costs do not rise dramatically for the attacker (Fig. 9.4) they do for the defender (Fig. 9.6).

It is assumed that if the benefit of avoiding eviction initially exceeds the estimate of the cost of a successful defence then the defender will resist. This is likely because costs will initially be predicted to be low (see above). However, because expected costs rise quickly but expected benefits remain the same (see above) there will come a time when expected costs exceed expected benefits and the defender will then give up. These changes may be plotted in a similar two dimensional state space as for the attacker. The prediction is that the higher the initial expected cost the shorter the fight, and the higher the expected benefit of victory the longer the fight. These are the predicted times for which the defender is prepared to resist, the actual fight might be shorter if the attacker gives up.

IS THE MODEL CONGRUENT WITH THE DATA?

The model generates fight times that in a relative sense are supported by the data. Thus attackers with much to gain fight for longer than do those with little to gain, and relatively large attackers are more persistent than relatively small attackers. Furthermore, defenders in good shells are more persistent (Elwood and Neil 1986b). What is required of the model, however, is that it generates realistic fight times in the absolute sense as well.

The model, as presented in Fig. 9.5, generates fight times for attackers between 2 and 62 seconds but real fights vary between about 2 and 600 seconds, with the occasional even longer fight. This apparent discrepancy is because the plots of changing estimates of V were derived from observations on crabs investigating shells, the apertures of which were blocked with cement, and median persistence values were used. Thus crabs in very poor shells investigating preferred size shells with blocked apertures persisted for an average of 66 seconds. There was, however, considerable variation in persistence times when investigating blocked shells, with some over 600 seconds (Fig. 9.7). It is interesting to note the remarkable similarity in the distribution of fight times in which the attacker gave up (Fig. 8.7b) and giving up times with blocked shells (Fig. 9.7). It seems that variation in each is similar and thus the model is a good predictor of the times for which attackers should be prepared to fight. If we assume that the benefits of retaining shells vary between 5 and 15 units of fitness, then expected defence durations will vary between 30 and 600 seconds, which again matches the observed durations (Fig. 8.7a). However, the important point is not that arbitrary fitness units produce predicted fight times similar to actual fight times, rather that the model works in a qualitative way.

COMPARISON WITH OTHER MODELS

The assumption that aggressive interactions may be determined primarily by two major factors has a long history. The conflict hypothesis proposes that

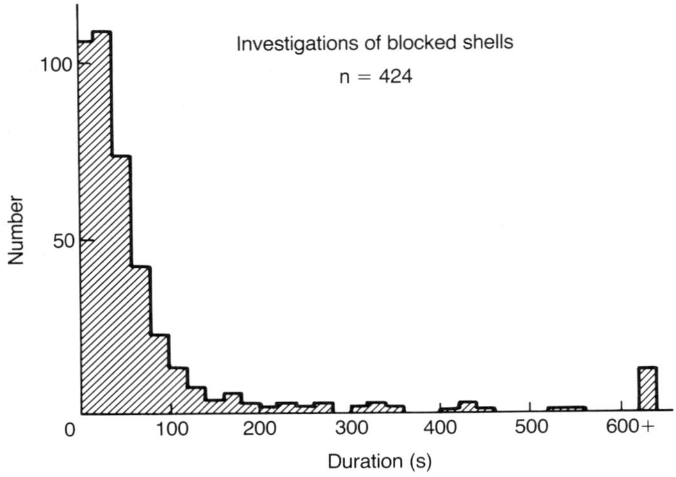

Figure 9.7 Distribution of the durations of investigation of blocked shells. Note the similarity with the distribution of attackers giving up (Fig 9.4 lower graph).

an interaction between the motivational systems controlling attacking and fleeing produces various aspects of aggressive displays and fights (reviewed by Baerends 1975). In particular, the conflict hypothesis has been used to explain differences between closely related species in aggressive behaviour (e.g. McKinney 1961). Recently the hypothesis has been developed to explain the activities observed during spider contests (Maynard Smith and Riechert 1984). Basically this study assumed that if the aggressive tendency exceeded the fleeing tendency the animal would be prepared to fight, whereas if the two tendencies were in balance it would be more likely to display. To produce a model that replicated observed behaviour, in particular transitions between specific acts, a number of assumptions had to be made concerning: (i) temporary components of motivational state that rise and then fall during a contest, and (ii) components that change as information is gathered. One major component of the model measured the size difference between the contestants and the other measured the value of the resource (i.e. estimates of *K* and *V* respectively). The predictions of the model were made more explicit by the use of a two-dimensional state space as shown in Fig. 9.8. The vectors in this model were labelled 'fear' and 'aggression' but the terms were used for 'convenience' (Riechert and Maynard Smith 1989). They could have been termed 'estimate of cost based on relative RHP' and 'estimate of the value of the resource'; the model would be essentially the same.

Riechert and Maynard Smith (1989) investigated the genetic basis of their

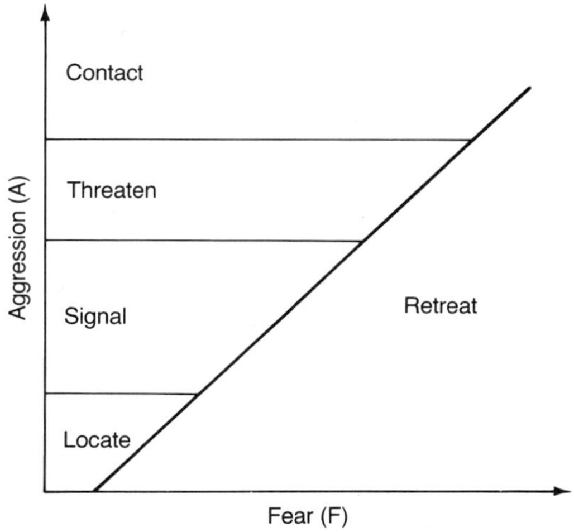

Figure 9.8 A model of the dynamics of territorial encounters between spiders (simplified from Maynard Smith and Riechert 1984), showing the rules that relate overt behaviour to levels of aggression and fear. (From Huntingford and Turner 1987.)

two-tendency model by examining the behaviour of two conspecific spider populations, their crosses and backcrosses. One population, from a grassland habitat, had more costly fights (in duration, number of acts performed, and risk of injury) than did a second population, from a river bank habitat. In the grasslands good web-sites are infrequent and hence of greater value to the owner (Riechert 1979, 1981). A curious finding, however, is that fights between F_1 hybrids are more costly and, in particular, more likely to lead to injury or death than fights between two grassland individuals. This is explained by the two factors of 'aggression' and 'fear' having a separate genetic basis. 'High aggression' is thought to be dominant to 'low aggression' and 'low fear' dominant to 'high fear'. The grassland population is thought to be homozygous for high aggression (dominant) and high fear (recessive) whereas the river bank population is homozygous for low aggression (recessive) and low fear (dominant). The F_1 is thus heterozygous with one gene dominant for high aggression and one dominant for low fear, which leads to costly fights. Riechert and Maynard Smith (1989) conclude 'that the genetic mechanisms underlying these traits associated with competitive success are sufficiently complex as not to constrain this spider from achieving those local competitive equilibria dictated by cost/benefit and frequency-dependent criteria. In other words, natural selection has a suitable template to operate on. This study offers a significant advance in our understanding of the decision-making systems controlling aggressive behaviour. It offers a rare insight into genetic influences on how animals make assessments. Those spiders with one or more dominant genes for 'high aggression' appear to assess the value of the resource more highly than those with double recessives for 'low aggression'. Furthermore, the key element in 'low fear' seems to be a devalued assessment of risk or cost.

In the model presented above to describe motivational change during hermit crab fights the estimate of cost was assumed to be relatively error free (Fig. 9.4). Note was made, however, of situations in which the estimate of cost was incorrect due to a paucity of information at one time and subsequently corrected as more information became available. This is precisely the situation modelled in the sequential assessment game of Enquist and Leimar (1983, 1987). This game models contests between pairs of opponents that differ in their fighting abilities, i.e. their abilities to inflict costs on the opponent and to avoid paying costs. The main assumption is that at the start of a contest opponents have poor information about each other's fighting ability. As the contest proceeds the opponents will assess, i.e. gather information about each other's fighting ability. Information is assumed to accumulate in a manner similar to statistical sampling (Enquist and Leimar 1983). There is assumed to be random error in an animal's estimate of its opponent's relative fighting ability but that this error is reduced as the interaction progresses. That is, there are increased numbers of opportunities to assess the opponent as the contest progresses so the 'sample size' of the estimate increases. This estimate is used continuously during the contest for strategic decisions about continuing or giving up. The problem for the animal is how to interpret each estimate, given that it contains random error. The intuitive answer is that the animal should place less reliance on early estimates than on later estimates. At which stage in the contest should the animal regard the estimates as reliable? The answer to this is mathematically complex (Enquist and Leimar 1983) but the solution, in terms of the ESS, is easily represented graphically. In Fig. 9.9 the vertical axis represents each animal's estimate of its fighting ability relative to that of its opponent and the horizontal axis represents the passage of time in the contest. There are two lines indicating how each contestant's estimate changes with time. The upper line indicates an animal that estimates that it has the greater fighting ability, the lower line an opponent that estimates it has the lower fighting ability. The changes in estimate vary as the number of samples increases i.e. as information is gathered. The approach of this model is novel because it introduces the concept of a switching line (Chapter 1) into the calculation of the ESS. Indeed the ESS is essentially the shape of the switching line within a motivational state space. In Fig. 9.9 this is shown as the lower curve. When the estimate of an animal's relative fighting ability falls below this switching line it will stop fighting.

An important determinant of the shape of the switching line is the value

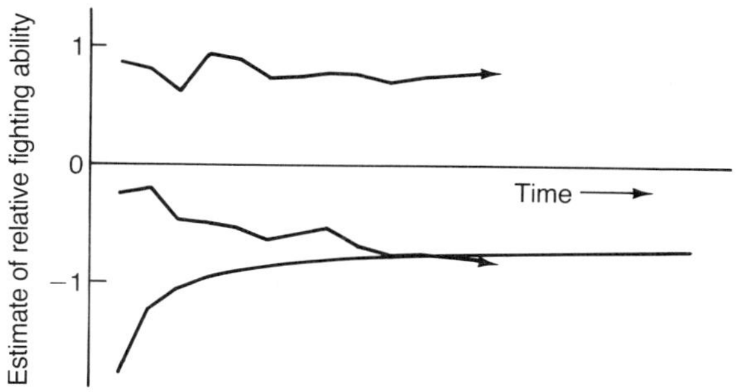

Figure 9.9 A causal factor space with time of fighting along the horizontal axis and estimate of relative fighting ability along the vertical axis. A stable switching line for a symmetrical situation is shown (both contestants use the same switching line). The two trajectories illustrate each contestant's estimate of its relative fighting ability during a contest. An animal continues to fight as long as the estimate is above the switching line. (From Leimar 1988.)

the animal places on the resource. The greater the value of the resource the lower will be the switching line (Fig. 9.10). Thus in the example shown in Fig. 9.10 of fights between spiders (Leimar 1988), fights between males for high-value females ($V = 40$) are predicted to be longer than those for low-value females ($V = 3$). This is because the switching line determines the level of reliability that the animal will accept. Those with low-value resources will accept low levels of reliability, and will stop fighting if there

Figure 9.10 The intruder's switching line (solid line) is an ESS for contests between 'typical' owners and intruders, both having resource value V = 13 eggs. The three broken lines are optimal switching lines against an intruder for owners with resource values V = 40, 10 and 3 eggs, respectively. See text for further explanations. (From Leimar 1988.)

is only a slight indication that they are inferior to the opponent. The model of the sequential assessment game takes into account situations in which either a single activity or different activities are performed (normally starting with activities of low cost but low information) (Leimar 1988; Enquist *et al.* 1990).

The sequential assessment model is important because it attempts to integrate functional explanations with causal explanations. The model, however, makes predictions similar to those made by the asymmetrical war of attrition game with incomplete information. The main difference is that in the war of attrition model a fight will continue until $K > V$ but in the sequential assessment game it continues until one of the animals assesses that K will exceed V for itself sooner than it will for the other animal. Both models predict longer fights when RHP is similar for the two contestants and longer fights when V is high. It is thus difficult to discriminate empirically between the models without having perfect knowledge of the costs of fights and of the value of resources. Good estimates of these are available from studies of fights between spiders (Austad 1983) and data from that study have been re-analysed to test the sequential assessment game (Leimar 1988). The data generally fit both models, but even with the excellent estimates available for K and V it is not possible to discriminate between the two models.

Although the models have yet to be discriminated empirically it is becoming increasingly obvious that animals do gather information during the course of a fight, that the information does result in motivational change, and that this motivational change influences strategic decisions. Thus the sequential assessment game offers a considerable advancement in the quest for reality in functional models as it takes into account the assessment abilities of animals.

It is clear that the models of Maynard Smith and Riechert (1984), Enquist and Leimar (1983) and that presented here have similar goals. Each attempts to incorporate assessments made during animal contests into the decision-making process. That of Enquist and Leimar (1983) is a general model using an ESS approach that could be applied to many contests, and has already been used to analyse fights between fish (Enquist *et al.* 1990), spiders (Leimar 1988) and insects (Englund and Olsson 1990). On the other hand the motivational models of Maynard Smith and Riechert (1984) and of hermit crab fights presented here are specific, although both have features in common, particularly the use of a two-dimensional state space. Similar models could thus be applied to other animal contests. The use of the two main factors of an estimate of resource value and an estimate of the cost of the fight should enable the integration of functional and motivational approaches to the study of animal contests. This we hope will result in an advance in our appreciation of how animals make assessments and decisions.

SUMMARY

A two-dimensional state space model is developed to explain motivational change during aggressive encounters of hermit crabs. The dimensions are (i) assessment of the cost of the fight, and (ii) assessment of resource value. The former is determined by relative size of the contestants and the point in time of the assessment; the latter by information available to each crab about the shells multiplied by the probability of obtaining or retaining the defender's shell. This probability will change during the course of the fight. As these assessments change during the fight so the point in motivational state will change. The trajectory of this point will determine fight duration and those generated by the model are similar to observed fight times. The motivational model is compared with functional models of aggression and with other models that examine motivational change (Enquist and Leimar 1983; Maynard Smith and Riechert 1984).

References

Abrams, P. (1978) Shell selection and utilization in a terrestrial hermit crab, *Coenobita compressus* (H. Milne Edwards). *Oecologia*, **34**, 239–53.
Abrams, P. (1980) Resource partitioning and interspecific competition in a tropical hermit crab community. *Oecologia*, **46**, 365–79.
Abrams, P. (1981a) Alternative methods of measuring competition applied to two Australian hermit crabs. *Oecologia*, **51**, 233–39.
Abrams, P. (1981b) Competition in an Indo-Pacific hermit crab community. *Oecologia*, **51**, 240–49.
Abrams, P. (1981c) Shell fighting and competition between two hermit crab species in Panama. *Oecologia*, **51**, 84–90.
Abrams, P. (1982a) Frequencies of interspecific shell exchanges between hermit crabs. *J. Exp. Mar. Biol. Ecol.*, **61**, 99–109.
Abrams, P. (1982b) Intraspecific shell exchange in the hermit crab *Clibanarius virescens* (Krauss). *J. Exp. Mar. Biol. Ecol.*, **59**, 89–101.
Adams, J. and Greenwood, P.J. (1983) Why are males bigger than females in pre-copula pairs of *Gammarus pulex*? *Behav. Ecol. Sociobiol.*, **13**, 239–41.
Adams, J. and Greenwood, P.J. (1987) Loading constraints, sexual selection and assortative mating in Percarid Crustacea. *J. Zool. Lond.*, **211**, 35–46.
Agassiz, A. (1875) Instinct? in hermit crabs. *Amer. J. Sci.*, **10**, 290–91.
Agassiz, A. (1876) How hermit crabs get possession of their shells. *Pop. Sci. Rev.*, **15**, 183–84.
Allee, W.C. and Douglis, M.B. (1945) A dominance order in the hermit crab *Pagurus bernhardus* Say. *Ecology*, **26**, 411–12.
Ameyaw-Akumfi, D. (1975) The breeding biology of two sympatric species of tropical intertidal hermit crabs, *Clibanarius chapini* and *C. senegalensis*. *Mar. Biol.*, **29**, 15–28.
Arak, A. (1988) Callers and satellites in the natterjack toad: evolutionarily stable decision rules. *Anim. Behav.*, **36**, 416–32.
Assem, J. van den. (1971) Some experiments on sex ratio and sex regulation in the Pteromalid *Lariophagus distinguendus*. *Neth. J. Zool.*, **20**, 329–52.
Austad, S.N. (1983) A game theoretical interpretation of male combat in the bowl and doily spider (*Frontinella pyramitela*). *Anim. Behav.*, **31**, 59–73.
Bach, C., Hazlett, B. and Rittschof, D. (1976) Effects of interspecific competition on fitness of the hermit crab *Clibanarius tricolor*. *Ecology*, **57**, 579–86.
Baerends, G.P. (1975) An evaluation of the conflict hypothesis as an explanatory principle for the evolution of displays, in *Function and Evolution of Behaviour*, (eds G.P. Baerends, C. Beer and A. Manning), Clarendon Press, Oxford, pp. 187–228.
Baerends, G.P., Brouwer, R. and Waterbole, H.T. (1955) Ethological studies on

Lebistes reticulatus (Peters): I. An analysis of the male courtship pattern. *Behaviour*, **8**, 249–334.

Balasch, J. and Mengual, V. (1974) The behavior of *Dardanus arrosor* in association with *Calliactis parasitica* in artificial habitat. *Mar. Behav. Physiol.*, **2**, 251–60.

Ball, E.E. Jr. (1968) Activity patterns and retinal pigment migration in *Pagurus* (Decapoda, Paguridae). *Crustaceana*, **14**, 302–306.

Ball, E.E. (1972) Observations on the biology of the hermit crab, *Coenobita compressus* H. Milne Edwards (Decapoda; Anomura) on the west coast of the Americas. *Rev. Biol. Trop.*, **20**, 265–73.

Ball, E.E. Jr. and Haig, J. (1972) Hermit crabs from eastern New Guinea. *Pac. Sci.*, **26**, 87–107.

Ball, E.E. and Haig, J. (1974) Hermit crabs from the tropical eastern Pacific. I. Distribution, color, and natural history of some common shallow-water species. *Bull. S. Calif. Acad. Sci.*, **73**, 95–104.

Barron L.C. and Hazlett, B.A. (1989) Directed currents: a hydrodynamic display in hermit crabs. *Marine Behav. Physiol.*, **15**, 83–87.

Berman, C., Francis, R.L. and Grant, W.C. Jr. (1973) Supplementary notes on the ecology and behavior of hermit crabs in the Frenchman Bay area. *Bull. Mt. Desert Is. Biol. Lab.*, **12**, **16**.

Bertness, M.D. (1980) Shell preference and utilization patterns in littoral hermit crabs of the Bay of Panama. *J. Exp. Mar. Biol. Ecol*, **48**, 1–16.

Bertness, M.D. (1981a) Competitive dynamics of a tropical hermit crab assemblage. *Ecology*, **62**, 751–61.

Bertness, M.D. (1981b) Conflicting advantages in resource utilization: the hermit crab housing dilemma. *Am. Nat.*, **118**, 432–37.

Bertness, M.D. (1981c) The influence of shell-type on hermit crab growth rate and clutch size (Decapoda, Anomura). *Crustaceana*, **40**, 197–205.

Bertness, M.D. (1981d) Interference, exploitation, and sexual components of competition in a tropical hermit crab assemblage. *J. Exp. Mar. Biol. Ecol.*, **49**, 189–202.

Bertness, M.D. (1981e) Pattern and plasticity in tropical hermit crab growth and reproduction. *Am. Nat.*, **117**, 754–73.

Bertness, M.D. (1981f) Predation, physical stress, and the organization of a tropical rocky intertidal hermit crab community. *Ecology*, **62**, 411–25.

Bertness, M.D. (1981g) Seasonality in tropical hermit crab reproduction in the Bay of Panama. *Biotropica*, **13**, 292–300.

Bertness, M.D. (1982) Shell utilization, predation pressure, and thermal stress in Panamanian hermit crabs: an interoceanic comparison. *J. Exp. Mar. Biol. Ecol.*, **64**, 159–87.

Bertness, M.D. and Cunningham, C. (1981) Crab shell-crushing predation and gastropod architectural defense. *J. Exp. Mar. Biol. Ecol.*, **50**, 213–30.

Bertness, M.D., Garrity, S.D. and Levings, S.C. (1981) Predation pressure and gastropod foraging: a tropical-temperate comparison. *Evolution*, **35**, 995–1007.

Birkhead, T.R. and Clarkson, K. (1980) Mate selection and precopulatory guarding in *Gammarus pulex*. *Z. Tierpsychol.*, **52**, 365–80.

Bishop, D.T., Cannings, C. and Maynard Smith, J. (1978) The war of attrition with random rewards. *J. Theor. Biol.*, **74**, 377–88.

Blackstone, N.W. (1984) The effects of history on the shell preference of the hermit crab *Pagurus longicarpus* (Say). *J. Exp. Mar. Biol. Ecol.*, **81**, 225–34.

Blackstone, N.W. (1985) The effects of shell size and shape on growth and form in the hermit crab *Pagurus longicarpus*. *Biol. Bull.*, **168**, 75–90.

Blackstone, N.W. and Joslyn, A.R. (1984) Utilization and preference for the introduced gastropod *Littorina littorea* (L.) by the hermit crab *Pagurus longicarpus* (Say) at Guilford, Connecticut. *J. Exp. Mar. Biol. Ecol.*, **80**, 1–10.

Bliss, D.E. (1968) Transition from water to land in decapod crustaceans. *Am. Zool.*, **8**, 355–92.

Bollay, M. (1964) Distribution and utilization of gastropod shells by the hermit crabs *Pagurus samuelis*, *P. granosimanus*, and *P. hirsutiusculus* at Pacific Grove, California. *Veliger* (suppl.), **6**, 71–6.

Boltt, R.E. (1961) Antennary feeding of the hermit crab, *Diogenes brevirostris* Stimpson. *Nature*, **192**, 1099–100.

Bookhout, C.G. (1964) Salinity effects on the larval development of *Pagurus bernhardus* (L.) reared in the laboratory. *Ophelia*, **1**, 275–94.

Brightwell, L.R. (1951) Some experiments with the common hermit crab (*Eupagurus bernhardus*) Linn., and transparent univalve shells. *Proc. Zool. Soc. Lond.*, **121**, 279–83.

Brightwell, L.R. (1952) Further notes on the hermit crab *Eupagurus bernhardus* and associated animals. *Proc. Zool. Soc. Lond.*, **123**, 61–4.

Brockman, H.J. and Dawkins, R. (1979) Joint nesting in a digger wasp as an evolutionary stable preadaptation to social life. *Behaviour*, **71**, 203–45.

Brockman, H.J., Grafen, A. and Dawkins, R. (1979) Evolutionarily stable nesting strategy in a digger wasp. *J. Theor. Biol.*, **77**, 473–96.

Brooks, W.R. and Mariscal, R.N. (1985) Protection of the hermit crab *Pagurus pollicaris* Say from predators by hydroid-colonized shells. *J. Exp. Mar. Biol. Ecol.*, **87**, 111–18.

Brown, A.F. and Thompson, D.B.A. (1986) Parasite manipulation of host behaviour: acanthocephalus and shrimps in the laboratory. *J. Biol. Educ.*, **20**, 121–27.

Burgess, J.W. (1976) Social spiders. *Sci. Am.*, **234**, 100–6.

Burggren, W.W. and McMahon, B.R. (1981) Oxygen uptake during environmental temperature change in hermit crabs: adaptation to subtidal, intertidal and supratidal habitats. *Physiol. Zool.*, **54**, 325–33.

Butler, C.G. (1949) *The Honeybee*. Oxford University Press. Oxford.

Caine, E.A. (1975) Feeding and masticatory structures of selected Anomura (Crustacea). *J. Exp. Mar. Biol. Ecol.*, **18**, 277–301.

Caine, E.A. (1976) Relationship between diet and the gland filter of the gastric mill in hermit crabs (Decapoda, Paguridae). *Crustaceana* **31**, 312–13.

Caine, E.A. (1980) Adaptations of a species of hermit crab (Decapoda, Paguridae) inhabiting sessile worm tubes. *Crustaceana* **38**, 306–310.

Caldwell, R.L. (1979) Cavity occupation and defensive behaviour in the stomatopod *Gonodactylus festai*: Evidence for chemically mediated individual recognition. *Anim. Behav.*, **27**, 194–201.

Caldwell, R.L. and Dingle, J. (1979) The influence of size differential in the mantis shrimp, *Gonodactylus viridis*. *Behaviour*, **69**, 255–64.

Carpenter, A. and Logan, W.H. (1945) Hermits don't always live alone. *Nat. Hist.*, **54**, 286–87.

Chalmers, N. (1983) The development of social relationships, in *Animal Behaviour: 3, Genes, development and learning*, (eds T.R. Halliday and P.J.B. Slater), Blackwell Scientific Publications, Oxford.

Charnov, E.L. (1981) Sex reversal in *Pandalus borealis*: effect of a shrimp fishery? *Mar. Biol. Letters*, **2**, 53–57.

Charnov, E.L. (1982) *The theory of sex allocation*, Princeton University Press, Princeton.

Charnov, E.L., Los-Den Hartogh, R.L., Jones, W.T., and van den Assem, J. (1981) Sex ratio evolution in a variable environment. *Nature*, **289**, 27–33.
Chase, I.D., Weissburg M. and Dewitt, T.H. (1988) The vacancy chain process: a new mechanism of resource distribution in animals with application to hermit crabs. *Anim. Behav.*, **36**, 1265–74.
Childress, J.P. (1972) Behavioral ecology and fitness theory in a tropical hermit crab. *Ecology*, **53**, 960–64.
Christensen, H.E. (1967) Ecology of *Hydractinia echinata* (Fleming) (Hydroidea, Athecata) I. Feeding Biology. *Ophelia*, **4**, 245–75.
Clutton-Brock, T.H., Albon, S.D., Gibson, R.M. and Guiness, F.E. (1979) The logical stag: Adaptive aspects of fighting in red deer (*Cervus elaphus* L.). *Anim. Behav.*, **27**, 211–25.
Coffin, H.G. (1960) The ovulation, embryology, and developmental stages of the hermit crab *Pagurus samuelis* (Stimpson). *Walla Walla Coll. Publs. Dep. Biol. Sci.*, **25**, 1–30.
Cohen, S. and McFarland, D.J. (1979) Time-sharing as a mechanism for the control of behaviour sequences during the courtship of the three-spined stickleback (*Gasterosteus aculeatus*). *Anim. Behav.*, **27**, 270–83.
Conover, M.R. (1975) Prevention of shell burial as a benefit hermit crabs provide to their symbionts (Decapoda, Paguridae). *Crustaceana*, **29**, 311–13.
Conover, M.R. (1976) The influence of some symbionts on the shell-selection behaviour of the hermit crabs, *Pagurus pollicarus* and *P. longicarpus*. *Anim. Behav.*, **24**, 191–94.
Conover, M.R. (1978) The importance of various shell characteristics to the shell-selection behavior of hermit crabs. *J. Exp. Mar. Biol. Ecol.*, **32**, 131–42.
Conover, M.R. (1979) Effect of gastropod shell characteristics and hermit crabs on shell epifauna. *J. Exp. Mar. Biol. Ecol.*, **40**, 81–94.
Courchesne, E. and Barlow, G.W. (1971) Effect of isolation on components of aggressive and other behavior in the hermit crab *Pagurus samuelis*. *Z. vergl. Physiologie*, **75**, 32–48.
Courtney, S.P. and Parker, G.A. (1985) Mating behaviour of the tiger blue butterfly (*Tarucus theophrastus*): competitive mate-searching when not all females are captured. *Behav. Ecol. Sociobiol.*, **17**, 213–21.
Cowles, R.P. (1919) Habits of tropical Crustacea. III. Habits and reactions of hermit crabs associated with sea anemones. *Phillipine J. Sci.*, **15**, 81–89.
Cowles, R.P. (1920) The transplanting of sea anemones by hermit crabs. *Proc. Nat. Acad. Sci.*, **6**, 40–42.
Crawford S.S. and Colgan P.W. (1989) Motivational models of courtship in male three-spine sticklebacks (*Gasterosteus aculeatus*). *Behaviour*, **109**, 285–302.
Culshaw, A.D. and Broom, D.M. (1980) The imminence of behavioural change and the startle response of chicks. *Behaviour*, **73**, 64–76.
Currey, J.D. and Hughes, R.N. (1982) Strength of the dogwhelk *Nucella lapillus* and the winkle *Littorina littorea* from different habitats. *J. Anim. Ecol.*, **51**, 47–56.
Cutress, C.E. and Ross, D.M. (1969) The sea anemone *Calliactus tricolor* and its association with the hermit crab *Dardanus venosus*. *J. Zool.*, **158**, 225–52.
Davenport, J. (1972) Effects of size upon salinity tolerance and volume regulation in the hermit crab *Pagurus bernhardus*. *Mar. Biol.*, **17**, 222–27.
Davenport, J., Busschots, P.M.C.F. and Cawthorne, D.F. (1980) The influence of salinity on behaviour and oxygen uptake of the hermit crab *Pagurus bernhardus* L. *J. Mar. Biol. Ass. U.K.*, **60**, 127–34.
Davies, N.B. (1978) Territorial defence in the speckled wood butterfly (*Pararge*

aegeria) (Lep., Satyridae): The resident always wins. *Anim. Behav.*, **26**, 138–47.

Davies, N.B. (1982) Behaviour and competition for scarce resources, in *Current Problems in Sociobiology*, (eds King's College Sociobiology Group), Cambridge University Press, Cambridge.

Davies, N.B. and Halliday, T.R. (1977) Optimal mate selection in the toad. *Nature*, **269**, 56–58.

Davies, N.B. and Halliday, T.R. (1978) Deep croaks and fighting assessment in toads *Bufo bufo*. *Nature*, **274**, 683–85.

Dawkins, R. and Dawkins, M. (1973) Decisions and the uncertainty of behaviour. *Behaviour*, **45**, 83–103.

Dawkins, R. and Krebs, J.R. (1978) Animal signals: information or manipulation, in *Behavioural Ecology: An Evolutionary Approach*, (eds J.R. Krebs and N.B. Davies), Blackwell Scientific Publications, Oxford, pp. 282–309.

De Wilde, P.A.W.J. (1973) On the ecology of *Coenobita clypeatus* in Curacao. *Stud. Fauna Curacao*, **44**, 1–138.

Dick J.T.A. and Elwood R.W. (1989) Assessments and decisions during mate choice in *Gammarus pulex* (Amphipoda). Behaviour, **109**, 235–46.

Dick J.T.A. and Elwood R.W. (1990) Symmetrical assessment of female quality by male *Gammarus pulex* during struggles over precopula females (Amphipoda). *Anim. Behav.*, **40**, 877–83.

Dingle, H. (1969) A statistical and information analysis of aggressive communication in the mantis shrimp *Gonodactylus bredini* Manning. *Anim. Behav.*, **17**, 561–75.

Dingle, H. (1983) Strategies of agonistic behavior in Crustacea, in *Studies in Adaptation; the Behavior of Higher Crustacea*, (eds S. Rebach and D. W. Dunham), John Wiley and Sons, New York, pp. 85–111.

Do, T.T., Kajihara, T. and Ho, J.S. (1984) The history of *Pseudomyicola spinosus* (Raffaele and Monticelli, 1885) from the blue mussel, *Mytilus edulis gallopro-vincialis* in Tokyo Bay, with notes on the production of a typical male. *Bull. Ocean Res. Inst. Univ. Tokyo*, **17**, 1–65.

Dowds, B.M. (1983) Aggression in the hermit crab *Pagurus bernhardus*. PhD Thesis, The Queen's University of Belfast, 157 pp.

Dowds, B.M. and R.W. Elwood. (1983) Shell wars: assessment strategies and the timing of decisions in hermit crab shell fights. *Behaviour*, **85**, 1–24.

Dowds, B.M. and R.W. Elwood. (1985) Shell wars II: the influence of relative size on decisions made during hermit crab shell fights. *Anim. Behav.*, **33**, 649–56.

Dunham, D.W. (1978a) On contrast and communication efficiency in hermit crabs. *Crustaceana*, **35**, 50–54.

Dunham, D.W. (1978b) Effect of chela white on agonistic success in a Diogenid hermit crab (*Calcinus laevimanus*). *Mar. Behav. Physiol.*, **335**, 137–44.

Dunham, D.W. and Tierney, A.J. (1983) The communicative cost of crypsis in a hermit crab *Pagurus marshi*. *Anim. Behav.*, **31**, 783–87.

Elmhirst, R. (1947) Ecology of *Eupagurus bernhardus*, the hermit crab. *Report of the Scottish Marine Biological Association*, pp. 18–20.

Elwood, R.W. and Adams, P. (1990) How hermit crabs (*Pagurus bernhardus*) deal with obstructions in the apertures of shells. *Ir. Nat. J.*, **23**, 180–85.

Elwood, R.W. and Glass, C.W. (1981) Negotiation or aggression during shell fights of the hermit crab *Pagurus bernhardus*? *Anim. Behav.*, **29**, 1239–44.

Elwood, R.W. and Kennedy, H.F. (1988) Sex differences in shell preferences of the hermit crab *Pagurus bernhardus* L. *Ir. Nat. J.*, **22**, 436–40.

Elwood, R.W. and Neil, S.J. (1986a) Information and motivational change, in *Quantitative Models in Ethology*, (eds P.W. Colgan and R. Zayan), Privat, I.E.C., Toulouse, pp. 97–107.

Elwood, R.W. and Neil, S. J. (1986b) Asymmetric contests involving two resources. *J. theor. Biol.*, **120**, 237–49.

Elwood, R.W. and Stewart, A. (1985) The timing of decisions during shell investigation by the hermit crab *Pagurus bernhardus*. *Anim. Behav.*, **33**, 620–27.

Elwood, R.W. and Stewart, A. (1987) Reproduction in the littoral hermit crab *Pagurus bernhardus*. *Ir. Nat. J.*, **22**, 252–55.

Elwood, R.W., McClean, A. and Webb, L. (1979) The development of shell preferences by the hermit crab *Pagurus bernhardus*. *Anim. Behav.*, **27**, 940–46.

Elwood, R.W., Gibson, J. and Neil, S. (1987) The amorous *Gammarus*: size assortative mating in *G. pulex*. *Anim. Behav.*, **35**, 1–6.

Englund, G. Olsson. (1990) Fighting and assessment in the net-spinning caddis larva *Arctopsyche ladogensis*. *Anim. Behav.*, **39**, 55–62.

Enquist, M. and Leimar, O. (1983) Evolution of fighting behaviour: decision rules and assessment of relative strength. *J. Theor. Biol.*, **102**, 387–410.

Enquist, M. and Leimar, O. (1987) Evolution of fighting behaviour: the effect of variation in resource value. *J. Theor. Biol.*, **127**, 187–205.

Enquist, M., Plane, E. and Roed, J. (1985) Aggressive communication in fulmars (*Fulmarus glacialis*) competing for food. *Anim. Behav.*, **33**, 1007–20.

Enquist, M., Leimar, O., Ljungberg, T., Mallner, Y. and Segerdahl, N. (1990) Test of the sequential assessment game I: fighting in the cichlid fish *Nannacara anomala*. *Anim. Behav.*, **40**, 1–14.

Fotheringham, N. (1975) Structure of seasonal migrations of the littoral hermit crab *Clibanarius vittatus* (Bosc). *J. Exp. Mar. Biol. Ecol.*, **18**, 47–53.

Fotheringham, N. (1976a) Effects of shell stress on the growth of hermit crabs. *J. Exp. Mar. Biol. Ecol.*, **23**, 299–305.

Fotheringham, N. (1976b) Hermit crab shells as a limiting resource (Decapoda, Paguridae). *Crustaceana*, **31**, 193–99.

Fotheringham, N. (1976c) Population consequences of shell utilization by hermit crabs. *Ecology*, **57**, 570–78.

Fotheringham, N. (1980) Effects of shell utilization on reproductive patterns in tropical hermit crabs. *Mar. Biol.*, **55**, 287–93.

Fotheringham, N. and Bagnall, R.A. (1976) Seasonal variation in the occurrence of planktonic larvae of sympatric hermit crabs. *J. Exp. Mar. Biol. Ecol.*, **21**, 279–87.

Fox, H.M. (1965) Confirmation of old observations on the behaviour of a hermit crab and its commensal sea anemone. *Ann. Mag. Nat. Hist.*, Ser. **13**, **8**, 173–75.

Frey, D.F. and Pimental, R.A. (1978) Principal component analysis, in *Quantitative Ethology*, (ed P.W. Colgan), Wiley-Interscience, London.

Fricke, H. and Fricke, S. (1977) Monogamy and sex change by aggressive dominance in coral reef fish. *Nature*, **266**, 830–32.

Fry, D.P. (1980) The evolution of aggression and the level of selection controversy. *Aggressive Behaviour*, **6**, 69–89.

Gerlach, S.A., Ekstrom, D.K. and Eckardt, P.B. (1976) Filter feeding in the hermit crab *Pagurus bernhardus*. *Oecologia*, **24**, 257–64.

Gilchrist, S. (1984) Evaluation of hermit crab shell indices. *Mar. Biol.*, **78**, 147–51.

Giles, N. (1984) Development of overhead fright response in wild and predator-naive three-spined sticklebacks, *Gasterosteus aculeatus* L. *Anim. Behav.*, **32**, 276–79.

Giles, N. and Huntingford, F.A. (1984) Predation risk and interpopulation variation in antipredator behaviour in the three-spined stickleback, *Gasterosteus aculeatus* L. *Anim. Behav.*, **32**, 264–75.

Godin, J.J. and Rangeley R.W. (1989) Living in the fast lane: effects of cost of locomotion on foraging behaviour in juvenile Atlantic salmon. *Anim. Behav.*, **37**, 943–54.

Grafen, A. (1979) The hawk-dove game played between relatives. *Anim. Behav.*, **27**, 905–7.

Grant, W.C. Jr. (1963) Notes on the ecology and behavior of the hermit crab, *Pagurus acadianus*. *Ecology*, **44**, 767–71.

Grant, W.C. Jr. and Pontier, P.J. (1973) Fitness in the hermit crab *Pagurus acadianus* with reference to *Hydractinia echinata*. *Bull. Mt. Desert Mar. Biol. Lab.*, **13**, 50–53.

Grant, W.C. Jr. and Ulmer, K.M. (1974) Shell selection and aggressive behavior in two sympatric species of hermit crabs. *Biol. Bull.*, **146**, 32–43.

Green, R.F. (1982) Optimal foraging and sex ratio in parasitic wasps. *J. theor. Biol.*, **95**, 43–48.

Greenwood, J.G. (1972) The mouthparts and feeding behaviour of two species of hermit crabs. *J. Nat. Hist.*, **6**, 325–37.

Greenwood, P.J. and Adams, J. (1984) Sexual dimorphism in *Gammarus pulex*: the effect of current flow on precopula pair formation. *Freshwater Biology*, **14**, 203–209.

Gross, M.R. (1984) Sunfish, salmon, and the evolution of alternative reproductive strategies and tactics in fishes, in *Fish Reproduction: strategies and tactics*, (eds G.W. Potts and R. J. Wooton), Academic Press, London.

Gross, M.R. (1985) Disruptive selection for alternative life histories in salmon. *Nature*, **313**, 47–48.

Gruber, M. and Shoup, J. (1969) Crabs move ashore. *Sea Frontiers*, **15**, 364–75.

Haas, F. (1950) Hermit crabs in fossil snail shells in Bermuda. *Ecology*, **31**, 152.

Halliday, T.R. (1976) The libidinous newt. An analysis of variations in the sexual behaviour of the male smooth newt, *Triturus vulgaris*. *Anim. Behav.*, **24**, 398–414.

Halliday, T.R. (1977) The effect of experimental manipulation of breathing behaviour on the sexual behaviour of the smooth newt, *Triturus vulgaris*. *Anim. Behav.*, **25**, 39–45.

Halliday, T.R. and Sweatman, H.P.A. (1976) To breathe or not to breathe: the newt's problem. *Anim. Behav.*, **24**, 551–61.

Hamilton, W.D. (1964) The genetical evolution of social behaviour, II. *J. Theor. Biol.*, **7**, 17–52.

Hamilton, W.D. (1967) Extraordinary sex ratios. *Science*, **156**, 477–88.

Hammerstein, P. (1981) The role of asymmetries in animal contests. *Anim. Behav.*, **29**, 193–205.

Hammerstein, P. and Parker, G.A. (1982) The asymmetric war of attrition. *J. Theor. Biol.*, **96**, 647–82.

Harper, D.G.C. (1985) Pairing strategies and mate choice in female robins *Erithacus rubecula*. *Anim. Behav.*, **33**, 862–75.

Harvey, I.F. and Corbet, P.S. (1986) Territorial interactions between larvae of the dragonfly *Pyrrhosoma nymphala*: outcome of encounters. *Anim. Behav.*, **34**, 1550–61.

Hazlett, B.A. (1966a) The behavior of some deep-water hermit crabs (Decapoda: Paguridae) from the straits of Florida. *Bull. Mar. Sci.*, **16**, 76–92.

Hazlett, B.A. (1966b) Factors affecting the aggressive behavior of the hermit crab *Calcinus tibicen*. *Z. Tierpsychol.*, **23**, 655–71.

Hazlett, B.A. (1966c) Observations on the social behaviour of the land hermit crab *Coenobita clypeatus* (Herbst). *Ecology*, **47**, 316–17.

Hazlett, B.A. (1966d) Social behavior of the Paguridae and Diogenidae of Curacao. *Stud. Fauna Curacao*, **88**, 1–143.

Hazlett, B.A. (1967) Interspecific shell fighting between *Pagurus bernhardus* and *Pagurus cuanensis* (Decapoda, Paguridae). *Sarsia*, **29**, 215–20.

Hazlett, B.A. (1968a) Communicatory effect of body position in *Pagurus bernhardus* (L.) (Decapoda, Anomura). *Crustaceana*, **14**, 210–14.

Hazlett, B.A. (1968b) Dislodging behavior in European Pagurids. *Pubbl. Staz. Zool. Napoli*, **36**, 138–39.

Hazlett, B.A. (1968c) Effects of crowding on the agonistic behavior of the hermit crab *Pagurus bernhardus*. *Ecology*, **49**, 573–75.

Hazlett, B.A. (1968d) The phyletically irregular social behavior of *Diogenes pugilator* (Anomura, Paguridae). *Crustaceana*, **15**, 31–34.

Hazlett, B.A. (1968e) The sexual behavior of some European hermit crabs (Anomura: Paguridae). *Pubbl. Staz. Zool. Napoli*, **36**, 238–52.

Hazlett, B.A. (1968f) Size relationships and aggressive behavior in the hermit crab *Clibanarius vittatus*. *Z. Tierpsychol.*, **25**, 608–14.

Hazlett, B.A. (1968g) Stimuli involved in the feeding behavior of the hermit crab *Clibanarius vittatus* (Decapoda, Paguridea). *Crustaceana*, **15**, 305–11.

Hazlett, B.A. (1969a) Further investigations of the cheliped presentation display in *Pagurus bernhardus* (Decapoda, Anomura). *Crustaceana*, **17**, 31–34.

Hazlett, B.A. (1969b) 'Individual' recognition and agonistic behaviour in *Pagurus bernhardus*. *Nature*, **222**, 268–69.

Hazlett, B.A. (1969c) Stone fighting in the crab *Cancellus spongicola* (Decapoda, Anomura, Diogenidae). *Crustaceana*, **16**, 219–20.

Hazlett, B.A. (1970a) The effect of shell size and weight on the agonistic behavior of a hermit crab. *Z. Tierpsychol.*, **27**, 369–74.

Hazlett, B.A. (1970b) Interspecific shell fighting in three sympatric species of hermit crabs in Hawaii. *Pac. Sci.*, **24**, 472–82.

Hazlett, B.A. (1970c) Tactile stimuli in the social behavior of *Pagurus bernhardus* (Decapoda, Paguridae). *Behaviour*, **36**, 20–48.

Hazlett, B.A. (1971a) Influence of rearing conditions on initial shell entering behavior of a hermit crab (Decapoda, Paguridae). *Crustaceana*, **20**, 167–70.

Hazlett, B.A. (1971b) Non-visual functions of Crustacean eyestalk ganglia. *Z. Vergl. Physiol.*, **71**, 1–13.

Hazlett, B.A. (1972a) Shell fighting and sexual behavior in the hermit crab genera *Paguristes* and *Calcinus*, with comments on *Pagurus*. *Bull. Mar. Sci.*, **22**, 806–23.

Hazlett, B.A. (1972b) Stimulus characteristics of an agonistic display of the hermit crab (*Calcinus tibicen*). *Anim. Behav.*, **20**, 101–107.

Hazlett, B.A. (1978) Shell exchanges in hermit crabs: aggression, negotiation, or both? *Anim. Behav.*, **26**, 1278–79.

Hazlett, B.A. (1980a) Patterns of information flow in the hermit crab *Calcinus tibicen*. *Anim. Behav.*, **28**, 1024–32.

Hazlett, B.A. (1980b) Patterns of information flow in the hermit crab *Calcinus tibicen*. *Anim. Behav.*, **28**, 1024–1032.

Hazlett, B.A. (1981a) The behavioral ecology of hermit crabs. *Ann. Rev. Ecol. Syst.*, **12**, 1–22.

Hazlett, B.A. (1981b) Daily movements of the hermit crab *Clibanarius vittatus*. *Bull. Mar. Sci.*, **31**, 177–83.

Hazlett, B.A. (1982a) Chemical induction of visual orientation in the hermit crab *Clibanarius vittatus*. *Anim. Behav.*, **30**, 1259–60.

Hazlett, B.A. (1982b) Resource value and communication strategy in the hermit crab *Pagurus bernhardus* (L.). *Anim. Behav.*, **30**, 135–39.

Hazlett, B.A. (1983a) Interspecific negotiations: mutual gain in exchanges of a limiting resource. *Anim. Behav.*, **31**, 160–63.

Hazlett, B.A. (1983b) Parental behavior in decapod crustacea, in *Studies in Adaptation: The behavior of higher crustacea*, (eds S. Rebach and DW Dunham), John Wiley and Sons, New York. pp. 171–96.

Hazlett, B.A. (1984) Variation in the pattern of shell exchange among hermit crabs. *Anim. Behav.*, **32**, 934–35.

Hazlett, B.A. (1987) Information transfer during shell exchange in the hermit crab *Clibanarius antillensis*. *Anim. Behav.*, **35**, 218–26.

Hazlett, B.A. (1989) Shell exchanges in the hermit crab *Calcinus tibicen.Anim. Behav.*, **37**, 104–11.

Hazlett, B.A. and Baron, L.C. (1989) Influence of shells on mating behaviour in the hermit crab *Calcinus tibicen*. *Beh. Ecol. Sociobiol.*, **24**, 369–76.

Hazlett, B.A. and Bossert, W.H. (1965) A statistical analysis of the aggressive communications systems of some hermit crabs. *Anim. Behav.*, **13**, 357–73.

Hazlett, B.A. and Bossert, W.H. (1966) Additional observations on the communications systems of hermit crabs. *Anim. Behav.*, **14**, 546–49.

Hazlett, B.A. and Estabrook, G.F. (1974a) Examination of agonistic behaviour by character analysis I. The spider crab *Microphyrs bicornutus*. *Behaviour*, **48**, 131–44.

Hazlett, B.A. and Estabrook, G. (1974b) Examination of agonistic behavior by character analysis. II. hermit crabs. *Behaviour*, **49**, 88–110.

Hazlett, B.A. and Herrnkind, W. (1980) Orientation to shell events by the hermit crab *Clibanarius vittatus* (Bosc) (Decapoda, Paguridae). *Crustaceana*, **39**, 311–14.

Hazlett, B.A. and Provenzano Jr, A.J. (1965) Development of behavior in laboratory reared hermit crabs. *Bull. Mar. Sci.*, **15**, 616–33.

Hazlett, B.A., Rubenstein, D. and Rittschoff, D. (1975) Starvation, energy reserves and aggression in the crayfish *Orconectes virilis* (Hagen, 1870) (Decapoda, Cambaridae). *Crustaceana*, **28**, 11–16.

Helfman, G.S. (1977a) Agonistic behavior of the coconut crab, *Birgus latro* (L.). *Z. Tierpsychol.*, **43**, 425–38.

Helfman, G.S. (1977b) Copulatory behavior of the coconut or robber crab *Birgus latro* (L.) (Decapoda, Anomura, Paguridae, Coenobitidae). *Crustaceana*, **33**, 198–202.

Herreid, C.F. and Full, R.J. (1986a) Energetics of hermit crabs during locomotion: the cost of carrying a shell. *J. Exp. Biol.*, **120**, 297–308.

Herreid, C.F. and Full, R.J. (1986b) Locomotion of hermit crabs (*Coenobita compressus*) on beach and treadmill. *J. Exp. Biol.*, **120**, 283–96.

Hipeau-Jacquotte, R. (1984) A new concept in the evolution of the Copepoda: *Pachypygus gibber* (Notodelphyidae), a species with two breeding males. *Crustaceana*, Suppl. **7**, 60–67.

Holmberg, K., Edsman, L., and Klint, T. (1989) Female mate preferences and male attributes in mallard ducks, *Anas platyrhynchos*. *Anim. Behav.*, **38**, 1–7.

Howard, R.D. (1978) The evolution of mating strategies in bull frogs, *Rana catesbeiana*. *Evolution*, **32**, 850–71.

Hunte, W., Myers, R.A. and Doyle, R.W. (1985) Bayesian mating decisions in an amphipod *Gammarus lawrencianus* Bousfield. *Anim. Behav.*, **33**, 366–72.

Huntingford, F.A. (1976) An investigation of the territorial behaviour of the three spined stickleback (*Gasterosteus aculeatus*) using principal components analysis. *Anim. Behav.*, **24**, 822–34.

Huntingford, F.A. and Turner, A. (1987) *Animal Conflict*, Chapman and Hall, London.

Hyatt, G.W. (1983) Qualitative and quantitative dimensions of crustacean aggression, in *Studies in Adaptation: the Behavior of Higher Crustacea*, (eds S. Rebach and D.W. Dunham), J. Wiley and Sons, New York, pp. 113–39.

Imafuku, M. (1981) Tidal activity rhythm of the hermit crab *Pagurus geminus* (McLaughlin). *Publ. Seto Mar. Biol. Lab.*, **26**, 327–36.

Imafuku, M. (1983a) New shell acquisition in the hermit crab, *Pagurus geminus*. *J. Ethol.* (Japan Ethol. Soc.), **1**, 91–100.

Imafuku, M. (1983b) Two hermit crabs inhabiting a single gastropod shell. *J. Ethol.*, **1**, 113–14.

Imafuku, M. (1984) Quality of shells occupied by the hermit crab *Pagurus geminus*: how many hermit crabs are satisfied with their shells? *J. Ethol.*, **2**, 31–36.

Imafuku, M. (1986) Sexual discrimination in the hermit crab *Pagurus geminus*. *J. Ethol.*, **4**, 39–47.

Imafuku, M. (1989) Shell fights in the hermit crab *Pagurus gaminus*: effect of cheliped use and shell rapping. *J. Ethology*, **7**, 35–9.

Jackson, H.G. (1913) *Eupagurus*. *Proc. Trans. Liverpool Biol. Sc.*, **27**, 495–573.

Jackson, N.W. (1988) Information gathering and decision making during shell selection by the hermit crab *Pagurus bernhardus*. PhD thesis, The Queen's University of Belfast, 203 pp.

Jackson, N.W. and Elwood, R.W. (1989a) Memory of information gained during shell investigation by the hermit crab *Pagurus bernhardus*. *Anim. Behav.*, **37**, 529–34.

Jackson, N.W. and Elwood, R.W. (1989b) How animals make assessments: information gathering by the hermit crab *Pagurus bernhardus*. *Anim. Behav.*, **38**, 951–57.

Jackson, N.W. and Elwood, R.W. (1990) Interrupting an assessment process to probe changes in the motivational state. *Anim. Behav.*, **39**, 1068–77.

Jensen, K. (1970) The interaction between *Pagurus bernhardus* (L.) and *Hydractinia echinata* (Fleming). *Ophelia*, **8**, 135–44.

Jensen, K. (1975) The profit to *Pagurus bernhardus* (L.) by the presence of *Hydractinia echinata* (Fleming). *Biokon. Reports*, **1**, 1–4.

Jensen, K. and Bender, K. (1973) Invertebrates associated with snail shells inhabited by *Pagurus bernhardus* (L.) (Decapoda). *Ophelia*, **10**, 185–92.

Kamalaveni, S. (1949) On the ovaries, copulation and egg formation in the hermit crab *Clibanarius olivaceous* Henderson (Crustacea, Decapoda). *J. Zool. Soc. India*, **1**, 120–28.

Karlson, R.H. and Cariolou, M.A. (1982) Hermit crab shell colonization by *Crepidula convexa* Say. *J. Exp. Mar. Biol. Ecol.*, **65**, 1–10.

Kellogg, C.W. (1976) Gastropod shells: a potentially limiting resource for hermit crabs. *J. Exp. Mar. Biol. Ecol.*, **22**, 101–11.

Kellogg, C.W. (1977) Coexistence in a hermit crab species ensemble. *Biol. Bull.*, **153**, 133–44.

Kinosita, H. and Okajima, A. (1968) Analysis of shell-searching behavior of the land hermit crab, *Coenobita rugosus* H. Milne Edwards. *J. Fac. Sci. Univ. Tokyo, Sect IV*, **11**, 293–358.

Klomp, H. and Teerink, B.J. (1962) Host selection and number of eggs per oviposition in the egg-parasitoid *Trichogramma embryophagum*. *Nature*, **195**, 1020–21.

Krebs, J.R. and McCleery, R.H. (1984) Optimization in behavioural ecology, in *Behavioural Ecology: An Evolutionary Approach*, (eds J.R. Krebs and N.B. Davies), Blackwell Scientific Publications: Oxford, pp. 91–121.

Kunze, J. and Anderson, D.T. (1979) Functional morphology of the mouthparts and gastric mill of the hermit crabs *Clibanarius taeniatis, C. virescens, Paguristes squamosus* and *Dardanus setifer* (Anomura, Paguridae). *Aust. J. Mar. Freshw. Res.*, **30**, 683–721.

Kuris, A.M. and Brody, M.S. (1976) Use of principal components analysis to describe the snail shell resource for hermit crabs. *J. Exp. Mar. Biol. Ecol.*, **22**, 69–77.

Kyriacou, C.P. and Hall, J.C. (1980) Circadian rhythm mutations in *Drosophila melanogaster* affect short-term fluctuations in the male's courtship song. *Proc. Nat. Acad. Sci. U.S.A.*, **77**, 6729–33.

Lancaster, I. (1988) *Pagurus bernhardus* (L) – An introduction to the natural history of hermit crabs. *Field studies*, **7**, 189–238.

Lancaster, I. (1990) Reproduction and life history strategy of the hermit crab *Pagurus bernhardus. J. Mar. Biol. Ass. UK*, **70**, 129–42.

Larkin, S. and McFarland, D.J. (1978) The cost of changing from one activity to another. *Anim. Behav.*, **26**, 1237–46.

Leimar, O. (1988) Evolutionary analysis of animal fighting. Ph.D. Dissertation, Stockholm University, 174 pp.

Leimar, O. and Enquist, M. (1984) Effects of asymmetries in owner-intruder conflicts. *J. Theor. Biol.*, **111**, 475–91.

Lindsey, B.G. and Grant, W.C. Jr. (1970) Aspects of niche diversification in two species of hermit crabs. *Bull. Mt. Desert Is. Biol. Lab.*, **10**, 39–41.

MacDonald, J.D., Pike, R.B. and Williamson, D.I. (1957) Larvae of the British species of *Diogenes, Pagurus, Anapagurus* and *Lithodes* (Crustacea, Decapoda). *Proc. Zool. Soc. Lond.* **128**, 209–58.

MacKay, D.C.G. (1945a) Left-right tendency in the hermit crab, *Calcinus herbsteii. J. Comp. Psychol.*, **38**, 131–33.

MacKay, D.C.G. (1945b) Gastropod shells inhabited by Hawaiian hermit crabs. *Nautilus*, **58**, 120–24.

Mainardi, D. and Rossi, A.C. (1969) Relations between social status and activity toward the sea anemone *Calliactis parasitica* in the hermit crab *Dardanus arrosor. Rend. Sci. Accad. Naz. Lincei*, Ser., **8**, **47**, 16–21.

Mainardi, D. and Rossi, A.C. (1971) L'influence d'experiences sociales recentes sur les rapports de dominance-soumission chez le pagure *Diogenes pugilator* (Anomura, Paguridea). *Rend. Sci. Accad. Naz. Lincei*, **51**, 389–404.

Manning, A. (1976) The place of genetics in the study of behaviour, in *Growing Points in Ethology*, (eds P.P.G. Bateson and R.A. Hinde), Cambridge University Press, Cambridge.

Markham, J.C. (1968) Notes on growth-patterns and shell-utilization of the hermit crab *Pagurus bernhardus* (L.). *Ophelia*, **5**, 189–205.

Markham, J.C. (1977) Preliminary note on the ecology of *Calcinus verrilli*, an endemic Bermuda hermit crab occupying attached vermetid shells. *J. Zool. Lond.*, **181**, 131–36.

Markham, J.C. (1978) Bopyrid isopods parasitizing hermit crabs in the northwestern Atlantic Ocean. *Bull. Mar. Sci.*, **28**, 102–17.

Mather, J.A. (1982) Choice and competition: their effects on occupancy of shell homes by *Octopus joubini. Mar. Behav. Physiol.*, **8**, 285–93.

Matthews, D.C. (1956a) Observations on ova fixation in the hermit crab *Eupagurus prideauxii. Publ. Staz. Napoli*, **31**, 248–63.

Matthews, D.C. (1956b) The probable method of fertilization in terrestrial hermit crabs based on a comparative study of spermatophores. *Pac. Sci.*, **10**, 303–9.

Maynard Smith, J. (1974) The theory of games and the evolution of animal conflict. *J. Theor. Biol.*, **47**, 209–21.

Maynard Smith, J. (1976) Evolution and the theory of games. *Amer. Sci.*, **64**, 41–45.

Maynard Smith, J. and Parker, G.A. (1976) The logic of asymmetric contests. *Anim. Behav.*, **24**, 159–75.

Maynard Smith, J. and Price, G.R. (1973) The logic of animal conflict. *Nature*, **246**, 15–18.

Maynard Smith, J. and Riechert, S.E. (1984) A conflicting-tendency model of spider agonistic behaviour: hybrid-pure population line comparisons. *Anim. Behav.*, **32**, 564–78.

McCleery, R.H. (1978) Optimal behaviour sequences and decision making, in *Behavioural Ecology: An Evolutionary Approach*, (eds J.R. Krebs and N.B. Davies), Blackwell Scientific Publications, Oxford.

McFarland, D.J. (1966) The role of attention in the disinhibition of displacement activities. *Q.J. Exp. Psychol*, **18**, 19–30.

McFarland, D.J. (1969) Mechanisms of behavioural disinhibition. *Anim. Behav.*, **17**, 238–42.

McFarland, D.J. (1971) *Feedback Mechanisms in Animal Behaviour*, Academic Press, London.

McFarland, D.J. and L'Angellier, A.B. (1966) Disinhibition of drinking during satiation of feeding behaviour in the Barbary dove. *Anim. Behav.*, **14**, 463–67.

McFarland, D. and Houston, A. (1981) *Quantitative Ethology: The State Space Approach*, Pitman Advanced Publishing Program, London.

McFarland, D.J. and Lloyd, I.H. (1973) Time-shared feeding and drinking. *Q.J. Exp. Psychol.*, **25**, 48–61.

McFarland, D.J. and Sibly, R. (1972) 'Unitary drives' revisited. *Anim. Behav.*, **20**, 548–63.

McFarland, D.J. and Sibly, R. (1975) The behavioural final common path. *Phil Trans. Roy. Soc. B*, **270**, 265–93.

McKinney, F. (1961) An analysis of the displays of the European Eider *Somateria mollissima mollissima* (Linnaeus) and the Pacific Eider *Somateria mollissima v. nigra* (Bonaparte). *Behaviour*, Suppl. **7**, 1–124.

McLaughlin, P.A. and Bailey-Brock, J.H. (1975) A new Hawaiian hermit crab of the genus *Trizopagurus* (Crustacea, Decapoda, Diogenidae), with notes on its behavior. *Pac. Sci.*, **29**, 259–66.

McLean, R.B. (1974) Direct shell acquisitions by hermit crabs from gastropods. *Experientia*, **30**, 206–8.

McLean, R. (1983) Gastropod shells: a dynamic resource that helps shape benthic community structure. *J. Exp. Mar. Biol. Ecol.*, **69**, 151–74.

McLean, R.B. and Mariscal, R.N. (1973) Protection of a hermit crab by its symbiotic sea anemone *Calliactis tricolor*. *Experientia*, **29**, 128–30.

McMahan, B.R. and Burggren, W.W. (1979) Respiration and adaptation to the terrestrial habitat in the land hermit crab *Coenobita clypeatus*. *J. Exp. Biol.*, **79**, 265–82.

Mercando, N.A. and Lytle, C.F. (1980) Specificity in the association between *Hydractinia echinata* and sympatric species of hermit crabs. *Biol. Bull.*, **159**, 337–48.

Mesce, K.A. (1982) Calcium-bearing objects elicit shell selection behavior in a hermit crab. *Science*, **215**, 993–95.

Metcalf, N.B., Huntingford, F.A. and Thorpe, J.E. (1987) The influence of predation risk on the feeding motivation and foraging strategy of juvenile Atlantic salmon. *Anim. Behav.*, **35**, 901–11.

Milinski, M. (1984) A predator's cost of overcoming the confusion-effect of swarming prey. *Anim. Behav.*, **32**, 1157–62.
Milinski, M. and Heller, R. (1978) Influence of a predator on the optimal foraging behaviour of sticklebacks (*Gasterosteus aculeatus* L.). *Nature Lond.*, **275**, 642–44.
Mitchell, K.A. (1973) Activities of two British species of *Pagurus* (Crustacea, Decapoda, Paguroidea). *Mar. Behav. Physiol.*, **2**, 229–36.
Mitchell, K.A. (1975) An analysis of shell occupation by two sympatric species of hermit crab. I. Ecological factors. *Biol. Bull.*, **149**, 205–13.
Mitchell, K.A. (1976) Shell selection in the hermit crab *Pagurus bernhardus*. *Mar. Biol.*, **35**, 335–43.
Moorhouse, J.E., Fosbrooke, I.H.M. and Ludlow, A.R. (1987) Stopping a walking locust with sound: an analysis of variation in behavioural threshold. *Exp. Biol.*, **46**, 193–201.
Neil, S.J. (1983) Contests for space in breeding *Cichlasoma meeki*: the role of resource holding potential. *Behaviour*, **87**, 270–82.
Neil, S.J. (1985) Size assessment and cues: studies of hermit crab contests. *Behaviour*, **92**, 22–38.
Neil, S.J. and Elwood, R.W. (1985) Behavioural modifications during egg-brooding in the hermit crab, *Pagurus bernhardus* L. *J. Exp. Mar. Biol. Ecol.*, **94**, 99–114.
Neil, S.J. and Elwood, R.W. (1986) Factors influencing shell investigation in the hermit crab, *Pagurus bernhardus*. *Ethology*, **73**, 225–34.
Niggemann, R. (1968) Zur Biologie und Okologie des Landeinsiedlerkrebses *Coenobita scaevola* Forskal am Roten Meer. *Oecologia*, **1**, 236–64.
Orians, G.H. and King, C.F. (1964) Shell selection and invasion rates of some Pacific hermit crabs. *Pac. Sci.*, **18**, 297–306.
Orton, J.H. (1927) On the mode of feeding of the hermit crab *Eupagurus bernhardus* and some other decapods. *J. Mar. Biol. Assoc. U.K.*, **14**, 909–21.
Otronen, M. (1984) Male contests for territories and females in the fly *Dryomyza anilis*. *Anim. Behav.*, **32**, 891–98.
Parker, G.A. (1970) Sperm competition and its evolutionary effect on copula duration in the dung fly *Scatophaga stercoraria*. *J. Insect Physiol.*, **16**, 1301–28.
Parker, G.A. (1974) Assessment strategy and the evolution of fighting behaviour. *J. Theor. Biol.*, **47**, 223–43.
Parker, G.A. (1984) Evolutionarily stable strategies, in *Behavioural Ecology: An Evolutionary Approach*, (eds J.R. Krebs and N.B. Davies), Blackwell, London, pp. 30–61.
Parker, G.A. and Rubenstein, D.I. (1981) Role assessment, reserve strategy, and acquisition of information in asymmetric animal contests. *Anim. Behav.*, **29**, 221–40.
Parker, G.A. and Stuart, R.A. (1976) Animal behaviour as a strategy optimiser: evolution of resource assessment strategies and optimal emigration thresholds. *Am. Nat.*, **110**, 1055–76.
Partridge, B.L. (1980) Background camouflage: an additional parameter in hermit crab shell selection and subsequent behavior. *Bull. Mar. Sci.*, **30**, 914–16.
Partridge, L. (1983) Genetics and Behaviour, in *Animal Behaviour 3: Genes, Development and Learning*, (eds T.R. Halliday and P.J.B. Slater), Blackwell Scientific Publications, Oxford.
Patton, W.K. and Robertson, D.R. (1980) Pair formation in a coral-inhabiting hermit crab. *Oecologia*, **47**, 267–69.
Perrill, S.A., Gerhardt, H.C. and Daniel, R.E. (1982) Mating strategy shifts in male

tree frogs (*Hyla cinerea*): an experimental study. *Anim. Behav.*, 30, 43–48.

Pike, R.B. and Williamson, D.I. (1959) Observations on the distribution and breeding of British hermit crabs and the stone crab (Crustacea: Diogenidae, Paguridae and Lithodidae). *Proc. Zool. Soc. Lond.*, **132**, 551–67.

Poulton, M.J. and Thompson, D.J. (1987) The effects of the acanthocephalan parasite *Pomphorhynchus laevis* on mate choice in *Gammarus pulex*. *Anim. Behav.*, **35**, 1577–88.

Provenzano, A.J. Jr. (1960) Notes on Bermuda hermit crabs (Crustacea; Anomura). *Bull. Mar. Sci. Gulf & Caribbean*, **10**, 117–24.

Provenzano, A.J. Jr. (1968) *Lithopagurus yucatanicus*, a new genus and species of hermit crab with a distinctive larva. *Bull. Mar. Sci.*, **18**, 627–44.

Provenzano, A.J. Jr. (1978) Larval development of the hermit crab *Paguristes spinips* Milne Edwards 1880 (Decapod, Diogenidae) reared in the laboratory. *Bull. Mar. Sci.*, **28**, 512–26.

Provenzano, A.J. Jr. and Rice, A.L. (1966) Juvenile morphology and the development of taxonomic characters in *Paguristes sericeus* H. Milne Edwards (Decapoda, Diogenidae). *Crustaceana*, **10**, 53–69.

Radinovsky, S. and Henderson, A. (1974) The shell game. *Nat. Hist.*, **83**, 22–29.

Rand, A.S. and Rand, W.M. (1976) Agonistic behaviour in nesting iguanas: a stochastic analysis of dispute settlement dominated by the minimisation of energy cost. *Z. Tierpsychol.*, **40**, 279–99.

Randall, J.E. (1964) Contribution to the biology of the queen conch, *Strombus gigas*. *Bull. Mar. Sci.*, **14**, 246–95.

Rebach, S. (1968) Orientation and movements of the hermit crab *Pagurus longicarpus*. *Bull. Ecol. Soc. Amer.*, **49**, 112–13.

Rebach, S. (1974) Burying behavior in relation to substrate and temperature in the hermit crab *Pagurus longicarpus*. *Ecology*, **55**, 195–98.

Rebach, S. (1978) The role of celestial cues in short range migrations of the hermit crab, *Pagurus longicarpus*. *Anim. Behav.*, **26**, 835–42.

Rebach, S. (1981) Use of multiple cues in short-range migrations of Crustacea. *Am. Midl. Nat.*, **105**, 168–80.

Rebach, S. and Dunham, D.W., (eds) (1983) *Studies in Adaptation: the Behavior of Higher Crustacea*, John Wiley & Sons, New York.

Reese, E.S. (1961) Inter- and intraspecific dominance relationships of sympatric species of intertidal hermit crabs. *Am. Zool.*, **1**, 382.

Reese, E.S. (1962a) Shell selection behaviour of hermit crabs. *Anim. Behav.*, **10**, 347–60.

Reese, E.S. (1962b) Submissive posture as an adaptation to aggressive behavior in hermit crabs. *Z. Tierpsychol.*, **19**, 645–51.

Reese, E.S. (1963) The behavioral mechanisms underlying shell selection by hermit crabs. *Behaviour*, **21**, 78–126.

Reese, E.S. (1968a) Annual breeding seasons of three sympatric species of tropical intertidal hermit crabs, with a discussion of factors controlling breeding. *J. Exp. Mar. Biol. Ecol.*, **2**, 308–18.

Reese, E.S. (1968b) Shell use: an adaptation for emigration from the sea by the coconut crab. *Science*, **161**, 385–86.

Reese, E.S. (1969) Behavioral adaptations of intertidal hermit crabs. *Am. Zool.*, **9**, 343–55.

Reese, E.S. (1983) Evolution, neuroethology, and behavioral adaptations of crustacean appendages, in *Studies in Adaptation: the Behavior of Higher Crustacea*, (eds S. Rebach and D.W. Dunham), John Wiley & Sons, New York, pp. 57–84.

Reese, E.S. and Kinzie, R.A. III. (1968) The larval development of the coconut or robber crab *Birgus latro* (L.) in the laboratory (Anomura, Paguridea). *Crustaceana* Suppl., **2**, 117–44.

Reyne, A. (1939) On the food habits of the coconut crab (*Birgus latro* L.), with notes on its distribution. *Arch. Neerland. Zool.*, **3**, 282–320.

Riechert, S.E. (1978) Games spiders play: Behavioural variability in territorial disputes. *Behav. Ecol. Sociobiol.*, **3**, 135–62.

Riechert, S.E. (1979) Games spiders play. II Resource assessment strategies. *Behav. Ecol. Sociobiol.*, **6**, 121–25.

Riechert, S.E. (1981) The consequences of being territorial: spiders, a case study. *Am. Nat.*, **117**, 871–92.

Riechert, S.E. (1984) Games spiders play. III Cues underlying context associated changes in agonistic behaviour. *Anim. Behav.*, **32**, 1–15.

Riechert, S.E. and Maynard Smith, J. (1989) Genetic analyses of two behavioural traits linked to individual fitness in the desert spider, *Agelenopsis aperta*. *Anim. Behav.*, **37**, 624–37.

Rittschof, D. (1980a) Chemical attraction of hermit crabs and other attendants to simulated gastropod predation sites. *J. Chem. Ecol.*, **6**, 103–18.

Rittschof, D. (1980b) Enzymatic production of small molecules attracting hermit crabs to simulated gastropod predation sites. *J. Chem. Ecol,.* **6**, 665–76.

Roberts, M.H. Jr. (1968) Functional morphology of mouth parts of the hermit crabs, *Pagurus longicarpus* and *P. pollicaris*. *Chesapeake Sci.*, **9**, 9–20.

Roberts, M.H. Jr. (1970) Larval development of *Pagurus longicarpus* Say reared in the laboratory, I. Description of larval instars. *Biol. Bull.*, **139**, 188–202.

Roberts, M.H. Jr. (1971a) Larval development of *Pagurus longicarpus* Say reared in the laboratory, II. Effects of reduced salinity on larval development. *Biol. Bull.*, **140**, 104–16.

Roberts, M.H. Jr. (1971b) Larval development of *Pagurus longicarpus* Say reared in the laboratory, III. Behavioral responses to salinity discontinuities. *Biol. Bull.*, **140**, 489–501.

Ross, D.M. (1960) The association between the hermit crab *Eupagurus bernhardus* (L.) and the sea anemone *Calliactis parasitica* Couch. *Proc. Zool. Soc. Lond.*, **134**, 43–57.

Ross, D.M. (1967) Behavioural and ecological relationships between sea anemones and other invertebrates. *Oceanogr. Mar. Biol. Ann. Rev.*, **5**, 291–316.

Ross, D.M. (1970) The commensal association of *Calliactis polypus* and the hermit crab *Dardanus gemmatus* in Hawaii. *Can. J. Zool.*, **48**, 351–57.

Ross, D.M. (1971) Protection of hermit crabs (*Dardanus* spp.) from octopus by commensal sea anemones (*Calliactis* spp.). *Nature*, **230**, 401–2.

Ross, D.M. (1975) The behavior of pagurids in symbiotic association with actinians in Japan. *Publ. Seto Mar. Biol. Lab.*, **22**, 157–70.

Ross, D. M. (1979) 'Stealing' of the symbiotic anemone, *Calliactis parasitica*, in intraspecific and interspecific encounters of three species of Mediterranean pagurids. *Can. J. Zool.*, **57**, 1181–89.

Ross, D.M. and Sutton, L. (1961a) The association between the hermit crab *Dardanus arrosor* (Herbst) and the sea anemone *Calliactis parasitica* (Couch). *Proc. R. Soc. Lond. Ser. B*, **155**, 282–91.

Ross, D.M. and Sutton, L. (1961b) The response of the sea anemone *Calliactis parasitica* to shells of the hermit crab *Pagurus bernhardus*. *Proc. R. Soc. Lond. Ser. B*, **155**, 266–81.

Ross, D.M. and Sutton, L. (1968) Detachment of sea anemones by commensal hermit crabs and by mechanical and electrical stimuli. *Nature*, **217**, 380–81.

Ross, D.M. and von Boletzky, S. (1979) The association between the pagurid *Dardanus arrosor* and the actinian *Calliactis parasitica*. Recovery of activity in 'inactive' *D. arrosor* in the presence of cephalopods. *Mar. Behav. Physiol.*, **6**, 175–84.

Ross, D.M. and Wada, T. (1975) The behavior of actinians in symbiotic association with pagurids in Japan. *Can. J. Zool.*, **53**, 1735–48.

Rossi, A.C. (1971) Dominance-subordinance relationships in the hermit crab *Diogenes pugilator* (Anomura, Paguridea). *Rev. Comp. Animal*, **5**, 153–62.

Rowe, G.T. and Menzies, R.J. (1968) Orientation in two bathyal, benthic decapods, *Munida valida* Smith and *Parapagurus pilosimanus* Smith. *Limnol. Oceanog.*, **13**, 549–52.

Rowell, C.H.F. (1961) Displacement grooming in the chaffinch. *Anim. Behav.*, **9**, 38–63.

Rubenstein, D.I. and Hazlett, B.A. (1974) Examination of the agonistic behaviour of the crayfish *Orconectes virilis* by character analysis. *Behaviour*, **50**, 193–216.

Rutherford, J.C. (1977) Removal of living snails from their shells by a hermit crab. *Veliger*, **19**, 438–39.

Sabourin, T.D. and Stickle, W.B. (1980) Respiration and osmoregulation responses of the hermit crab *Clibinarius vittatus* in response to salinity changes. *J. Exp. Mar. Biol. Ecol.*, **46**, 241–54.

Salt, G. (1937) The sense used by *Trichogramma* to distinguish between parasitised and unparasitised hosts. *Proc. R. Soc. (Series B)*, **122**, 57–75.

Samuelson, T.J. (1970) The biology of six species of Anomura (Crustacea, Decapoda) from Raunefjorden, Western Norway. *Sarsia*, **45**, 25–52.

Schembri, P.J. (1982) Feeding behaviour of fifteen species of hermit crabs (Crustacea: Decapoda: Anomura) from the Otago region, southeastern New Zealand. *J. Nat. Hist.*, **16**, 859–78.

Schiejsma, K. (1935) Observations on *Hydractinia echinata* (Fleming) and *Eupagurus bernhardus* (L.). *Arch. Neerlandaises Zool.*, **1**, 261–314.

Schmidt, J.M. and Smith, J.J.B. (1987) Short interval time measurement by a parasitoid wasp. *Science*, **237**, 903–905.

Schumacher, H. (1977) A hermit crab, sessile on corals, exclusively feeds by feathered antennae. *Oecologia*, **27**, 371–74.

Scully, E.P. (1978) Utilization of surface foam as a food source by the hermit crab, *Pagurus longicarpus* Say, 1817. *Mar. Behav. Physiol.*, **5**, 159–62.

Scully, E.P. (1979) The effects of gastropod shell availability and habitat characteristics on shell utilization by the intertidal hermit crab *Pagurus longicarpus* Say. *J. Exp. Mar. Biol. Ecol.*, **37**, 139–52.

Scully, E.P. (1983a) The behavioral ecology of competition and resource utilization among hermit crabs, in *Studies in Adaptation: the Behaviour of Higher Crustacea*, (eds S. Rebach and D.W. Dunham), J. Wiley & Sons, New York, pp. 23–55.

Scully, E.P. (1983b) The effects of shell availability on intraspecific competition in experimental populations of the hermit crab, *Pagurus longicarpus* Say. *J. Exp. Mar. Biol. Ecol.*, **71**, 221–36.

Scully, E.P. (1986) Shell investigation behavior of the intertidal hermit crab *Pagurus longicarpus* Say. *J. Crustacean Biology*, **6**, 749–56.

Seeley, T.D. (1985) *Honeybee Ecology*. Princeton University Press, Princeton, 201 pp.

Selbie, C.M. (1921) The Decapoda Reptantia of the coast of Ireland. Part II. Paguridae. *Scient. Invest. Fish. Brch. Ireland*, **1**, 1–68.

Shoup, J.B. (1968) Shell opening by crabs of the genus Calappa. *Science*, **160**, 887–88.

Shumway, S.E. (1978) Osmotic balance and respiration in the hermit crab, *Pagurus bernhardus*, exposed to fluctuating salinities. *J. Mar. Biol. Ass. UK*, **58**, 869–76.

Sibly, R. and McCleery, R.H. (1976) The dominance boundary method of determining motivational state. *Anim. Behav.*, **24**, 108–24.

Sibly, R. and McFarland, D.J. (1976) On the fitness of behaviour sequences. *Amer. Nat.*, **110**, 601–17.

Sigurjonsdottir, H. and Parker, G.A. (1981) Dung fly struggles: Evidence for assessment strategy. *Behav. Ecol. Sociobiol.*, **8**, 219–30.

Simbolotti, G., Putters, F.A. and van den Assem, J. (1987) Rates of attack and control of the offspring sex ratio in the parasitic wasp *Lariophagus distinguendus* in an environment where host quality varies. *Behaviour*, **100**, 1–32.

Slack-Smith, S.M. (1982) Anemone shells. *Australian Shell News*, **40**, 2.

Slater, P.J.B. (1983) The development of individual behaviour, in *Animal Behaviour: 3, Genes, Development and Learning*, (eds T.R. Halliday and P.J.B. Slater), Blackwell Scientific Publications, Oxford.

Smart, J.L. (1983) Undernutrition, maternal behaviour, and pup development, in *Parental Behaviour of Rodents*, (ed R.W. Elwood), John Wiley & Sons Ltd., Chichester.

Snow, P. (1973) The antennular activities of the hermit crab *Pagurus alaskensis* (Benedict). *J. Exp. Biol.*, **58**, 745–65.

Snyder-Conn, E. (1980) Tidal clustering and dispersal of the hermit crab *Clibanarius digueti*. *Mar. Behav. Physiol.*, **7**, 135–54.

Snyder-Conn, E.K. (1981) The adaptive significance of clustering in the hermit crab *Clibanarius digueti*. *Mar. Behav. Physiol.*, **8**, 43–53.

Spight, T.M. (1977) Availability and use of shells by intertidal hermit crabs. *Biol. Bull.*, **152**, 120–33.

Sripathi, K.S. Ajmal Khan and Natarajan, R. (1977) Shell selection by the land hermit crab *Coenobita cavipes* Stimpson. *Indian J. Mar. Sci.*, **6**, 163–65.

Stachowitsch, M. (1977) The hermit crab microbiocoenosis – the role of mobile secondary hard bottom elements in a north Adriatic benthic community, in *Biology of Benthic Organisms*, (eds B.F. Keegan, P.O. Ceidigh and P.J.S. Boaden), Pergamon Press, London, pp. 549–58.

Stachowitsch, M. (1979) Movement, activity pattern, and role of a hermit crab population in a sublittoral epifaunal community. *J. Exp. Mar. Biol. Ecol.*, **39**, 135–50.

Stachowitsch, M. (1980) The epibiotic and endolithic species associated with the gastropod shells inhabited by the hermit crabs *Paguristes oculatus* and *Pagurus cuanensis*. *Mar. Ecol.*, **1**, 73–101.

Taylor, P.R. (1979) An association between an amphipod, *Liljeborgia* sp., and the hermit crab, *Pagurus hemphilli* (Benedict). *Mar. Behav. Physiol.*, **6**, 185–88.

Taylor, P.R. (1981) Hermit crab fitness: the effect of shell condition and behavioral adaptations on environmental resistance. *J. Exp. Mar. Biol. Ecol.*, **52**, 205–18.

Telford, M. and Daxboeck, C. (1978) *Porcellana sayana* Leach (Crustacea: Anomura) symbiotic with *Strombus gigas* (Linnaeus) (Gastropoda: Strombidae) and with three species of hermit crabs (Anomura: Diogenidae) in Barbados. *Bull. Mar. Sci.*, **28**, 202–205.

Thompson, M.T. (1903) The metamorphoses of the hermit crab. *Proc. Boston Soc. Nat. Hist.*, **31**, 147–209.

Thornhill, R. (1984) Fighting and assessment in *Harpo bittacus* scorpionflies. *Evolution*, **38**, 204–14.

Toates, F. (1980) *Animal Behaviour: A Systems Approach*, John Wiley & Sons, Chichester.

Van Iersel, J.J.A. and Bol, A.C.A. (1958) Preening of two tern species. A study of displacement activities. *Behaviour*, **13**, 1–88.

Vance, R.R. (1972a) Competition and mechanism of coexistence in three sympatric species of intertidal hermit crabs. *Ecology*, **53**, 1062–74.

Vance, R.R. (1972b) The role of shell adequacy in behavioral interactions involving hermit crabs. *Ecology*, **53**, 1075–83.

Verrell, P.A. (1983) The influence of the ambient sex ratio and intermale competition on the sexual behaviour of the red-spotted newt, *Notophthalmus viridescens* (Amphibia: Urodela: Salamandridae). *Behav. Ecol. Sociobiol.*, **13**, 307–13.

Verrell, P.A. (1986) Wrestling in the red-spotted newt (*Notophthalmus viridescens*): resource value and contestant asymmetry determine contest duration and outcome. *Anim. Behav.*, **34**, 398–402.

Volker, L. (1967) Zur Gehausewahl des land-einsiedlerkrebses *Coenobita scaevola* Forskal vom Roten Meer. *J. Exp. Mar. Biol. Ecol.*, **1**, 168–90.

vom Saal, F.S. (1984) Proximate and ultimate causes of infanticide and parental behaviour in male house mice, in *Infanticide: Comparative and Evolutionary Perspectives*, (eds G. Hausfater and S.B. Hrdy), Aldine Press, New York.

Waage, J.K. and Lane, J.A. (1984) The reproductive strategy of a parasitic wasp. II. Sex allocation and local mate competition in *Trichogramma evanescens*. *J. Anim. Ecol.*, **53**, 417–26.

Waage, J.K. and Ng, S.M. (1984) The reproductive strategy of a parasitic wasp. I Optimal progeny and sex allocation in *Trichogramma evanescens*. *J. Anim. Ecol.*, **53**, 401–15.

Wang, D. and Jillson, D.A. (1979) On the relevance of small gastropod shells to competing hermit crab species. *Veliger*, **21**, 488–89.

Ward, P.I. (1983) Advantages and a disadvantage of large size for male *Gammarus pulex* (Crustacea: Amphipoda). *Behav. Ecol. Sociobiol.*, **14**, 69–76.

Ward, P.I. (1984) The effects of size on the mating decisions of *Gammarus pulex* (Crustacea, Amphipoda). *Z. Tierpsychol.*, **64**, 174–84.

Warner, R.R. (1975) The adaptive significance of sequential hermaphroditism in animals. *Am. Nat.*, **109**, 61–82.

Wenner, A.M. (1972) Sex ratio as a function of size in marine Crustacea. *Am. Nat.*, **106**, 321–50.

Werren, J.H. (1980) Sex ratio adaptations to local mate competition in a parasitic wasp. *Science*, **208**, 1157–59.

Werren, J.H. (1983) Sex ratio evolution under local mate competition in a parasitic wasp. *Evolution*, **37**, 116–24.

Wetterer, J.K. (1989) Mechanisms of prey choice by planktivorous fish: perceptual constraints and rules of thumb. *Anim. Behav.*, **37**, 955–67.

Wicksten, M.K. (1977) Shells inhabited by *Pagurus hirsutiusculus* (Dana) at Coyote Point Park, San Francisco Bay, California. *Veliger*, **19**, 445–46.

Wicksten, M.K. (1979) Digging by the hermit crab *Isocheles pilosus* (Holmes) (Decapoda, Anomura, Diogenidae). *Crustaceana*, Suppl., **5**, 100.

Wilber, T.P. Jr. and Herrnkind, W. (1982) Rate of new shell acquisition by hermit crabs in a salt marsh habitat. *J. Crust. Biol.*, **2**, 588–92.

Wilber, T.P. Jr. and Herrnkind, W. (1984) Predaceous gastropods regulate new-shell supply to salt marsh hermit crabs. *Mar. Biol.*, **79**, 145–50.

Wilcox, R.S. and Ruckdeschel, T. (1982) Food threshold territoriality in a water strider (*Gerris remigis*). *Behav. Ecol. Sociobiol.*, **11**, 85–90.

Wilson, D.P. (1949) Notes from the Plymouth aquarium. *J. Mar. Biol. Assoc. UK*, **29**, 345–52.

Wilson, E.O. (1971) *The Insect Societies*. Harvard University Press, Cambridge, Mass.

Wright, H.O. (1973) Effect of commensal hydroids on hermit crab competition in the littoral zone of Texas. *Nature*, **241**, 139–40.

Yamaguchi, S. (1938) Spawning and development of *Coenobita rugosus*. *Bull. Sci. Kyusu Univ.*, **8**, 163–75.

Young, A.M. (1978) Desiccation tolerances for three hermit crab species *Clibanarius vittatus* (Bosc), *Pagurus pollicaris* Say and *P. longicarpus* Say (Decapoda, Anomura) in the North Inlet Estuary, South Carolina, U.S.A. *Est. Coastal Mar. Sci.*, **6**, 117–22.

Young, A.M. (1979a) Differential utilization of gastropod shells by three hermit crab species in North Inlet, South Carolina, U.S.A. *Crustaceana*, Suppl., **5**, 101–104.

Young, A.M. (1979b) Osmoregulation in larvae of the striped hermit crab *Clibanarius vittatus* (Decapoda, Anomura, Diogenidae). *Est. Coastal Mar. Sci.*, **9**, 595–602.

Young, A.M. (1979c) Osmoregulation in three hermit crab species *Clibanarius vittatus*, *Pagurus longicarpus* and *P. pollicaris* (Crustacea; Decapoda, Anomura). *Comp. Biochem. Physiol.*, **63**A, 377–82.

Young, A.M. (1980) Temperature-salinity tolerances of three hermit crab species, *Clibanarius vittatus* (Bosc), *Pagurus longicarpus* Say and *P. pollicaris* Say (Crustacea: Decapoda; Anomura). *Ophelia*, **19**, 27–35.

Young, A.M. and Hazlett III, T.L. (1978) The effect of salinity and temperature on the larval development of *Clibanarius vittatus* (Bosc) (Crustacea; Decapoda; Diogenidae). *J. Exp. Mar. Biol. Ecol.*, **34**, 131–41.

Zipser, E. and Vermeij, G.J. (1978) Crushing behavior of tropical and temperate crabs. *J. Exp. Mar. Biol. Ecol.* **31**, 155–72.

Index